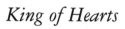

King of Hearts

KING

OF

HEARTS

The True Story of the
Maverick Who Pioneered
Open Heart Surgery

G. Wayne Miller

TIMES 𝕿 BOOKS

RANDOM HOUSE

Library of Congress Cataloging-in-Publication Data
Miller, G. Wayne.
King of hearts: the true story of the maverick who pioneered
open heart surgery / G. Wayne Miller.
p.   cm.
Includes index.
ISBN 0-8129-3003-7
1. Lillehei, C. Walton, 1918–1999.   2. Surgeons—
United States Biography.   3. Heart—Surgery—United States—
History.   I. Title.
RD598.M523   2000
617.412'092 [B]—dc21
99-38264

Random House website address: www.randomhouse.com
Printed in the United States of America on acid-free paper
24689753
FIRST EDITION
Book design by Susan Hood

Special Sales
Times Books are available at special discounts for bulk purchases
for sales promotions or premiums. Special editions, including
personalized covers, excerpts of existing books, and corporate
imprints, can be created in large quantities for special needs.
For more information, write to Special Markets, Times Books,
201 East 50th Street, New York, New York 10022, or call
800-800-3246.

For Joel P. Rawson, editor, mentor, and friend

This is entirely a work of nonfiction; it contains no composite characters or scenes, and no names have been changed. Nothing has been invented.

The author has used direct quotations only when he heard or saw (as in a letter) the words, and he paraphrased all other dialogue and statements—omitting quotation marks—once he was satisfied that these took place.

Contents

Introduction

Few of the great stories of medicine are as palpably dramatic as the invention of open heart surgery. Few triumphs came at such tragic cost.

I realized this early in writing *King of Hearts,* when I began to discover the countless people—mostly babies and young children—who died on the operating table after surgeons had opened them up, then been unable to fix their failing hearts. As a father, I understood the desperation that drove parents to entrust their little girl or boy to such a surgeon, who could promise nothing but a heroic effort; faced with terminal heart disease, almost anything was worth a try.

I understood less what motivated doctors to keep trying when so many children perished, literally in their hands. Indeed, many doctors dropped out—the human cost was too high, the emotional toll too devastating. But some persevered. Some, like C. Walton Lillehei, the protagonist of this story—the surgeon whom many call the Father of Open Heart Surgery—pushed ahead through all the bleeding and the dying until they finally got it right, only a few decades ago.

How far medicine has come. Today, cardiac surgery is nearly

as commonplace as removing an appendix, with almost one million open heart procedures and coronary bypasses—and more than 2,200 heart transplants—performed every year in the United States alone.

And yet, open heart surgery remains a great wonder.

While preparing this book, I watched Richard A. Jonas, chief heart surgeon at Boston's Children's Hospital, operate on a baby girl; only a week old, she was born with hypoplastic left heart syndrome, in which the entire left half of the heart is essentially missing. Not long ago, this baby would have died in a matter of hours or days—but Jonas saved her. After the operation, he delivered the news to the parents, whose four hours of waiting is something I hope never to experience. As the parents cried tears of joy, I thought of the countless times in hospitals around the world where this scene plays out every day, for patients young and old.

———

This is the story of a quest—a quest that some considered impossible, and others called murderous. Most of it occurred in the 1950s and early 1960s, when jet jockeys piloted the first rocket ships into space.

The best of the open heart pioneers shared much with the early astronauts. These young doctors had returned from war with ambition. They believed in breaking rules, in both their personal and professional lives. They did not recoil from death; they were beyond brave, existing in some rarefied place where taking risks is as fundamental as oxygen.

They were a unique group, these early open heart pioneers: some three dozen or so doctors whose collective contributions launched an infant science. Yet even in this elite group, Walt Lillehei (a Norwegian name, pronounced *Lilla-high*) stood apart.

After surviving war and then a case of deadly lymphatic cancer, which his mentor cut out of him in an operation that itself nearly killed him, Lillehei seemed to know no fear. He was supremely self-confident, of course, and he was a genius, driven by forces that he himself could never clearly articulate. Lillehei almost certainly lost more patients than any of his peers—yet he kept going, perhaps because he was unusually acquainted with death. For even after his cancer operation, Lillehei lived under a sentence: he had only a 25 percent chance of surviving five years.

Far from being embittered, Lillehei was a highly compassionate surgeon; without exception, everyone I interviewed who knew his bedside manner praised it. His young patients and their families idolized him.

Still, Lillehei was hardly a saint; a handsome man with piercing blue eyes, he lived his motto, "Work hard, play hard!" to the fullest—ultimately, with terrible consequences for him, his career, and his lovely wife.

I became interested in the story of open heart surgery while researching one of my previous books, *The Work of Human Hands,* set at Children's Hospital in Boston, where I met Craig Lillehei, Walt's son. Craig is a fine surgeon but a modest person, and he rarely mentioned his father. So while I knew C. Walton Lillehei had played some role in the development of cardiac surgery, I had no idea how important he'd been until he came to the Harvard Medical School to lecture in June of 1992. I sat through that lecture amazed, and then I went to the library.

There I found some scholarly books on the history of cardiac surgery, written in academic prose by physicians. A few journalists had also written books—but none (that I ever found, at least) went much beyond the scientific achievements to the story of desperate patients and families and their daring doctors, the foundation of my narrative.

I present *King of Hearts,* then, as both a history and a medical drama.

G. Wayne Miller
Pascoag, Rhode Island
October 1999

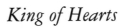

King of Hearts

WALT LILLEHEI, MEDICAL SCHOOL GRADUATE

Prologue:
Red Alert

ON THE DAY that some feared he crossed over into madness, the surgeon C. Walton Lillehei woke at his usual hour, six o'clock. He ate his ordinary light breakfast, read the morning paper, kissed his wife and three young children good-bye, then drove his flashy Buick convertible to University Hospital in Minneapolis. The first patients of the day were already unconscious when Lillehei dressed in scrubs and entered the main operating area. It was March 26, 1954.

Lillehei walked into Room II, where the doctors who would assist him were preparing two operating tables for the baby and the adult who would soon be there. Nearly all of the doctors were young—younger even than Lillehei, who was only thirty-five. Most were residents—surgeons still in training who were devoted to Lillehei not only because he was an outstanding surgeon, but also because he seemed to live for risk and he overflowed with unconventional new ideas.

Lillehei checked on the pump and the web of plastic tubes that would connect the adult to the baby. He confirmed that two teams of anesthesiologists were ready, that the OR supervisor had briefed the many nurses, and that the blood bank was steeled for possible massive transfusions.

He confirmed that the adult—who was the baby's father—had not changed his mind about their being the subjects of this experiment, which no doctor had attempted before.

It looks good, said Lillehei. I think we're ready to go.

———

Elsewhere in University Hospital, a nurse roused the baby.

Thirteen-month-old Gregory Glidden was an adorable boy with big ears and a fetching smile that had endeared him to the staff during the three straight months they had cared for him. He was unusually scrawny, but at the moment there was no other outward sign that he was sick. His appearance was deceiving. Gregory had been born with a hole between the lower chambers of his heart—a type of defect that no surgeon had ever been able to fix. In fact, Gregory was dying. Dr. Lillehei doubted he would last the year.

Unlike many nights in his short life, Gregory had slept well and he awoke in good spirits. The nurse cleaned his chest with an antibacterial solution and dressed him in a fresh gown, but she could not give him breakfast; for surgery, his body had to be pure. A resident administered penicillin and a preoperative sedative, and the baby became drowsy again. Then an orderly appeared and spoke softly to Gregory about the little trip he would be taking—that he would travel safe in his crib, with his favorite toys and stuffed animals.

One floor below his son, Lyman Glidden was also headed for surgery. His wife, Frances, had come by to see him off, and as they waited for Lyman to be wheeled away, they were thinking not only of Gregory. They were remembering their daughter LaDonnah, who had been born with the same defect as their little boy. Somehow LaDonnah had survived, in relatively good health, until the age of twelve. Then, in the spring of 1950, she became gravely ill, and one night that September, she died in her sleep.

The Gliddens could never forget finding her body, cold and rigid in her bed.

———

It was half past seven. In the operating room next to Lillehei, Chief of Surgery Owen H. Wangensteen was cutting into a woman he hoped to cure of cancer. Wangensteen had not checked on Lillehei, nor had he told the young surgeon yet of the ruckus that had developed yesterday afternoon, when another of University Hospital's powers—Chief of Medicine Cecil J. Watson, an internist often at odds with Wangensteen—had discovered what Lillehei intended to do.

Watson already knew that Lillehei had joined the quest to correct extreme defects inside the opened heart—a race that so far had produced only corpses, in Minneapolis and elsewhere. He knew of Lillehei's dog research—of how the surgeon and his young disciples regularly worked past midnight in their makeshift laboratory in the attic of a university building. He'd heard of Lillehei's new open heart technique, in which the circulatory systems of two dogs were connected with a pump and tubes; thus joined, the donor dog supported the life of the patient dog, enabling Lillehei to close off the vessels to the patient dog's heart, open the heart, and repair a life-threatening defect. But until yesterday afternoon, when today's operating schedule had been distributed, Watson had not known that Lillehei—with Wangensteen's blessing—was taking cross-circulation into the operating room.

This was madness!

Watson went to University Hospital's director, who alone had authority to stop an operation. The director summoned Wangensteen, and the three men had it out.

How could such an experiment be allowed? Watson demanded to know. For the first time in history, one operation had the potential to kill two people. Yet, paradoxically, how could the Gliddens refuse? They lived in the north woods of Minnesota, where Lyman worked the mines and Frances stayed home with their many children. They lacked the guidance of a human-experimentation committee, for none existed in 1954. They would never consult a lawyer, for they were willing to try almost anything to spare their baby their daughter's fate.

And was it any wonder that Wangensteen had blessed Lillehei? Of all the resplendent surgeons on Wangensteen's staff, Lillehei was unquestionably the crown prince—the most likely to bring the University of Minnesota a Nobel prize, which the chief of surgery all but craved. Blue-eyed and blond—a man who liked all-night jazz clubs and pretty women—brilliant Walt could do no wrong in Owen's eyes.

The chief of medicine was appalled. Had they forgotten the girl who had preceded Gregory Glidden in University Hospital's Room II, poor Patty Anderson, who'd been lost in a river of blood?

———

Gregory was brought into Room II and transferred to one of its two operating tables. His crib with his toys and stuffed animals was sent back into the hall and he was in the company of masked strangers. A doctor placed an endotracheal tube down his throat and turned the gas on.

Asleep, Gregory was stripped of his gown and left naked under the glare of hot lights. How small he was—smaller than a pillow, smaller than most laboratory dogs. His heart would be a trifle, his vessels thin as twine.

All set? Lillehei said.

The people with him were.

Lillehei washed Gregory's chest with surgical soap. With a scalpel, he cut left to right on a line just below the nipples.

Observers in Room II's balcony leaned forward for a better look. On the operating room floor, a crowd of interns and residents climbed up on stools.

Lillehei split the sternum, the bone that joins the ribs, and opened a window into Gregory with a retractor.

6

Nestled between his lungs, Gregory's plum-colored little heart came into view. It was noisy; with his hand, Lillehei felt an abnormal vibration.

Still, the outside anatomy appeared normal: the great vessels were in their proper places, with no unnatural connections between. So far, no surprises.

It looks okay, Lillehei said. You may bring in the father now.

PATTY ANDERSON, CHRISTMAS 1950

1

A River of Blood

LOYD ANDERSON returned to Minnesota, hoping to forget
what he'd seen on the battlefields of Europe, a decade before C. Walton Lillehei attempted what no surgeon had before. Lloyd's wife, Betty, needed to move on, too. A car crash
when she was a teen had put Betty in a coma, and she was different when she emerged, high-strung and fretful.

The end of World War Two promised new beginnings.
Lloyd the infantryman came home a father; Betty had become
pregnant while he was on leave, and their first child, Patricia
Lee, had arrived in his absence on April 7, 1945. At forty-six
hours, it had been a miserable labor, but mother and daughter
had left the hospital apparently healthy. Lloyd got a job in a
Minneapolis tractor factory; Betty was home with Baby. With
only an ordinary measure of luck, life at last would be sweet.

So maybe it was nothing when the Andersons began to notice that the blood vessels in their daughter's neck pulsated
strangely; maybe it was only their imagination that, as Patty
grew, her tiny chest seemed to bulge. New parents see the
worst in everything. Still, better not to take chances. Betty
took Patty, almost two, to the family doctor.

9

Listening to her heart, the doctor detected a murmur.

Patty deteriorated with astonishing speed after that. She caught frequent colds. She tired easily and experienced great difficulty sleeping. She was abnormally pale and thin—except for that bulge in her chest. That wasn't imagination. That kept getting bigger, as if something inside of Patty were desperate to escape.

In June 1947, a cardiologist at a University of Minnesota clinic diagnosed an atrial septal defect, or ASD—a hole in the wall that separates the heart's upper chambers. Cardiac diagnosis was unreliable in the 1940s, but if the doctor was correct, Patty was doomed. There was no fix for an ASD, no way to open the heart and sew up a hole without bleeding the patient to death. Like many of the fifty thousand or so other Americans born every year with such defects, Patty might not reach adulthood. She might not make first grade.

Three years passed.

Doctors put Patty on a low-salt diet and made her eat liver, rich in iron. She was given oxygen and digitalis, an herbal derivative that stimulates the heart—and one of the rare cardiac medications of any kind in those years. She underwent radiation therapy. But nothing worked. Patty's heart kept failing, and as a result, her lungs were also. Stairs exhausted her and she was consumed by thirst. So severe was her torment that she could sleep only by sitting up in bed. Taxed close to the breaking point, her heart was double its normal size. It was, quite literally, ready to burst.

And then the Andersons heard about Dr. Clarence Dennis, a University of Minnesota surgeon. Dennis was senior member of a staff that included such accomplished young doctors as C. Walton Lillehei.

For five years, Dennis had been developing a machine that would temporarily function as a patient's lungs and heart. With it, he hoped to shut off the vessels to a defective heart, open the heart, right its wrongs, then close it back up. Patty's parents listened to Dennis describe how his machine had worked on laboratory dogs. No one could guarantee its success in the human, of course, but Dennis was ready to try. All he lacked was a volunteer.

The Andersons did not answer immediately. For all the fancy language, the situation was brutally simple: Dennis wanted a human guinea pig. He wanted to hook their daughter up to a newfangled machine they'd never seen and, in a room they were forbidden to enter, entrust it with her life.

And yet, the Andersons' only child—their precious Patty— was slipping away. You could see it in her look, already marred by a terrible case of crossed eyes, and in her smile, which seemed forced and weary, as if she had never known joy. She did not play much with other children: she couldn't keep up and Betty was terrified of germs. Patty had been hospitalized three times already in her short life, most recently for a month and a half, including Christmas of 1950. Betty was chain-smoking and drinking coffee in dizzying quantities. Lloyd was revisiting battlefields in his sleep.

———

Clarence Dennis was born inventive. As a boy, he built a radio, a record player, and a lathe, and he was handy with Model Ts. Fascinated by the concept of free energy, he set out, at his basement workbench, to achieve the impossible. His father said to his mother, The boy doesn't have common sense. He's been working down there for weeks trying to make a perpetual-motion machine!

Dennis graduated magna cum laude from Harvard College and earned his M.D. with high honors at Johns Hopkins, then served his internship and residency at Minnesota. Impressed by Owen H. Wangensteen, the chief of surgery, Dennis decided to stay. Dennis was a professor when the chief asked him to join what many held to be surgery's greatest quest.

It was the autumn of 1945, when heart surgery consisted primarily of repair of peripheral structures such as the aorta, the vessel that sends freshly oxygenated blood back into the body. Fixing the worst defects—those inside the heart—would require time, room, and visibility, all of which were then unattainable. To open the living heart was to kill, in a river of blood that ran dry in less than a minute.

Only a handful of surgeons had yet embarked on the quest, and almost without exception they believed the answer would be found in a heart-lung machine: a mechanical apparatus that could sustain the patient's life while blood was detoured past the sick heart. The principle of supporting life by artificial means was not new; for decades, scientists had been able to keep tissue alive outside of the body with the assistance of various contraptions. The most publicized example was a sliver of embryonic chicken heart muscle that the great French surgeon and Nobel laureate Alexis Carrel managed to keep alive, with the help of a machine, for almost thirty years in his laboratory. Among those inspired by Carrel was aviator Charles Lindbergh, whose sister-in-law's heart disease prompted him, in 1929, to approach the French surgeon with the idea of building an artificial heart—an impossibility at that time, as Lindbergh soon learned.

Clarence Dennis's first inspiration was the artificial kidney, invented in Nazi-occupied Holland by the Dutch doctor Willem J. Kolff. Working in secret, Kolff built eight artificial kidneys out of wood, metal, and other materials that he scrounged from industrialist friends. All sixteen of Kolff's early dialysis patients died. "The first patient who truly recovered—most likely thanks to the artificial kidney—and who otherwise would have died, was number seventeen," Kolff later recalled. "She was a National Socialist, an old woman who most of my countrymen would have been most happy to send off to her final reward." Clarence Dennis thought the guts of Kolff's device (cellulose sausage casing, of all things) might properly oxygenate blood, a fundamental requirement of any heart-lung machine, but his laboratory studies disproved that hypothesis.

So Dennis traveled east to Philadelphia, home of surgeon John H. Gibbon Jr., who'd been working on a heart-lung machine for more than a decade, longer than anyone else. Gibbon was a guarded man, more so as his competition intensified, but in 1945, he welcomed a visitor from the Midwest.

Thank God you want to build one of your own, said Gibbon to Dennis. Now maybe people won't keep telling me I'm crazy!

Dennis returned to Minneapolis with Gibbon's blueprints, which he took to the university's machinists, who tooled and helped assemble the parts. The first tests used cow blood, available free from a slaughterhouse, but Dennis wasn't satisfied with that design, nor with several others. Years passed. Then one night Dennis woke up, literally from a dream, with the idea for another modification. A prototype was built and tested. Dennis was ready to try it on a living being, and because its heart is nearly identical to a human's, he chose a dog. He anesthetized the animal, split its chest, connected it to the machine, and opened its heart for half an hour. That dog lived, but the next thirty or so died.

Dennis made more modifications and ran a new series of tests. Lost dogs outnumbered survivors—because of mistakes, not a flaw in design, Dennis believed. The mistakes were instructive; once identified, they could be prevented.

I think we're ready to give this a try, said Dennis to surgeon Richard Varco, his principal partner.

No one else, not even John Gibbon, had ever tried a heart-lung machine on a human—no one had even demonstrated that the diseased human heart could be safely opened, never-mind tolerate surgery on its innermost reaches. A flight of stairs could nearly kill Patty Anderson—what of the trauma of scalpel and stitches, compounded by the steely intrusion of a machine?

But Dennis had toiled five long years in the lab. He was eager to try.

On the morning of April 6, 1951, the day before her sixth birthday, Patty Anderson was brought into Room II and anesthetized. Richard Varco opened her chest with a clean incision and pried her ribs apart; then, he and Dennis went to work on the mass of scar tissue that entombed her crippled heart. The scar tissue was the legacy of earlier, exploratory surgery that had resolved nothing.

For four hours the surgeons labored, isolating vital vessels and peeling away fibrous growth to expose the object of their effort. Patty's blood pressure dipped dangerously but they did not flinch. Far greater hazards lay ahead.

Finally, the girl's heart was freed.

Dennis and Varco had seen its shadowy picture on X ray, of course, and even a stranger couldn't help but notice how her chest bulged—and sometimes pounded frighteningly, almost like a drum being banged. Naked under hot lights, this heart was an unforgettable sight: grotesquely swollen and audibly quivering, like a creature from the depths of the sea. It was remarkable that Patty Anderson was alive.

Dennis's heart-lung machine fit on a table, but it was enormously complicated, with pumps, valves, switches, motors, a flow meter, a solenoid, a reservoir, and a series of slowly rotating stainless-steel disks onto which blood from the patient gently flowed (by removing carbon dioxide and infusing fresh oxygen, the disks would serve as lungs; the pumps would substitute for Patty's heart). Applying this technology to the patient was no simple matter, either. Sixteen people were needed: Dennis and Varco, the principal surgeons; two anesthesiologists; two assistant surgeons; two technicians; two nurses; a person to draw blood samples; a person in charge of transfusion; and four people on the machine.

At fourteen minutes past noon, Varco and Dennis began to connect Patty. Tubes were placed in the vessels that returned blood from throughout the body to the heart; they would deliver Patty's oxygen-depleted blood to the machine. Another tube, for return of freshly oxygenated blood from the machine, was placed into an artery on the other side of Patty's heart. The heart could now be tied off, its function usurped by man.

Dennis checked to be sure that everyone was ready.

Sixteen people were.

Pump on, Dennis said.

It was 1:22 P.M.

Varco cut into Patty's heart. It bled. A suction tube was applied and Varco cut deeper. The bleeding intensified. Varco opened the heart's upper right-hand chamber—the right atrium.

The machine was functioning flawlessly, but now the bleeding was overwhelming: blood poured into the heart and the suction was hopelessly inadequate. The surgeons could barely see—yet, between suctionings, they could make out enough to know they had bigger trouble than bleeding.

The diagnosis was wrong.

Patty's defect was no simple hole, but something neither surgeon had ever seen, not in an anatomy text nor in the pathology lab: a cluster of abnormalities involving holes in the heart's center, and two grossly misshapen valves. Varco felt around with his fingers, hoping for enlightenment.

Dear God, thought Dennis. We're like Columbus without a compass.

Now what?

They were ten minutes into it. They couldn't keep Patty on the machine forever, but the bleeding wouldn't let up.

Now what?

Varco grasped the edges of what seemed to be the worst hole and began suturing; eleven stitches and it was shut. But there was nothing they could do about the valves. Patty's heart, which had continued to beat, was beginning to slow. Dennis sewed the outer wall shut and signaled the pump off.

At 2:02 P.M., after forty minutes on the machine, Patty was disconnected.

Her own heart took over, but not willingly. Dennis massaged it by hand, and together with medications and transfusions kept it going, falteringly, for another half hour.

But all was futile now.

At 2:45 P.M., Dennis surrendered. Patty Anderson's heart stopped for good.

Dennis could not bear breaking the news to Betty and Lloyd Anderson; he let Varco, who knew them better. The Andersons were devastated, for despite all of the doctors' warnings, they had believed Patty would make it.

On what would have been her sixth birthday, April 7, 1951, the Andersons' only child was buried. A newspaper published a three-quarter-inch funeral notice, but there was no obituary, no photograph, no clue that a little girl with sadly crossed eyes had made history.

In analyzing what went wrong, Dennis was satisfied with the performance of his machine. Until diagnostic techniques improved, the risk of surprise would always be there; repair of an ASD was within reach, Dennis still believed.

And so he decided to try again, on two-year-old Sheryl L. Judge, whose heart had been failing since winter.

Word of the impending operation set University Hospital abuzz. A mob of observers, including C. Walton Lillehei, had gathered in Room II on the morning of May 31, 1951, when Dennis and Varco opened the child.

This time, luck was with them: the diagnosis was correct. Sheryl had an ASD, a simple hole.

They were about to begin the repair when Varco shouted: What's going on? You're blowing bubbles!

16

Air was leaking out of the coronary vessels. Air—deadly to all tissue because it blocks the flow of blood through life-sustaining capillaries.

Baffled, Dennis and Varco turned to the men on the machine.

The men on the machine looked horror-stricken.

They had let the reservoir run dry, and Sheryl had been pumped full of air. Brain, heart, liver—all poisoned by air.

It was an accident. One of the technicians was light-headed and feverish from a cold; when starting the machine, he'd forgotten to activate a fail-safe circuit that would have prevented the reservoir from emptying.

They reprimed the machine, but it was too late.

Eight hours after surgery, Sheryl Judge died.

As he watched from the back of Dennis's operating room, Walt Lillehei was absorbed. A junior member of University Hospital's surgical staff, Lillehei was beginning to think the heart was where he wanted to make his mark.

Whatever proved to be Lillehei's life's work, most who knew him believed he would succeed—with more than an ordinary splash, the betting was. Equally comfortable in the operating room or drinking martinis in a smoke-filled nightclub, Lillehei had already won a national award for his research. He was close to a professorship. He'd published a number of scientific articles and addressed national gatherings of surgeons—commanding audiences that had taken notice of the handsome young surgeon from Minnesota.

As accomplished as his peers were, Walt always seemed a step ahead. And what it boiled down to was this: Walt's brain was wired differently. Where others might debate whether the glass was half empty or half full, Walt would question whether a glass was the right vessel in the first place. And if he'd determined it wasn't, he already had a better idea.

Lillehei was impressed by Dennis's daring and the years of

17

meticulous study and testing that preceded clinical use of his heart-lung machine; Dennis had a fine mind, too. But the complexity of his creation troubled Lillehei. So many parts, so many possibilities for error, as Sheryl Judge had tragically demonstrated. And scoring sixteen people from the same sheet on such a delicate symphony—to Lillehei, it was preposterous.

You're doomed to failure with sixteen doctors, Lillehei thought. Sixteen doctors can't agree on anything!

———

Had you looked closely at Lillehei that May of 1951, you would have seen a scar near the thirty-two-year-old man's left ear. A surgeon evidently had been at him.

One morning not two years before, in late 1949, while he was nearing the end of his long training, Lillehei had found a bump on his face. A colleague had removed it and pronounced it either a benign tumor or a harmlessly enlarged lymph node. A specimen was sent to the University Hospital pathology department for microscopic analysis. This was routine.

The results stunned Owen Wangensteen, Lillehei's boss and mentor.

The pathologists reported that Lillehei had lymphosarcoma, or non-Hodgkin's lymphoma, one of the deadliest of cancers. Most of its victims died within five years of its onset.

There must be an error, thought Wangensteen. Lymphosarcoma is so rare. Walt is so healthy, and so young.

For the time being, Wangensteen kept the report a secret. He did not tell Lillehei, nor did Lillehei ask. Lillehei had accepted the initial diagnosis and gone about his business; bad news, he believed, would have found him immediately.

WALT LILLEHEI, ARMY OFFICER

2

Surgeon-Scientists

JENS KRISTIAN LILLEHEI left Tysnes, a heart-shaped island in Norway, in 1885, when he was seventeen. He was from a family of poor fishermen and farmers and, like so many others, he sailed to America seeking a better life. He got barely a taste. After becoming a bricklayer, marrying, and having two sons, Jens died of tuberculosis in 1898 in Minnesota at the age of thirty. His sons were not yet in school.

Jens's widow, Paluda, was determined that her children would make something of themselves. With two dollars a day cleaning houses, she put her sons through the University of Minnesota. The younger became a doctor. The older, Clarence, became a dentist and married Elizabeth Walton, a professional piano player. Clarence was in the army when the first of their three children, all sons, was born, on October 23, 1918.

Walt was a handsome boy, with Nordic features that one day would enchant women. He enjoyed the outdoors, especially the pond near his home in Edina, a suburb of Minneapolis. He played sports and he enjoyed winning, but off the field he was an easygoing boy—and a bit of a loner. Walt liked nice clothes. He liked the neighborhood library and the Edina Country Club, to which his parents belonged.

Despite their comfortable existence, the Lilleheis instilled in their children the value of hard work; Walt was not simply a Boy Scout, but an Eagle Scout—and as soon as he was old enough, he became a caddy. The Lilleheis also stressed independence. Accompanied by their dog Jiggs, a stray that showed up on the doorstep one day, Walt and his brothers would set off on Saturday mornings for the pond, not having to return until suppertime. And before they drove, the boys were free to take the trolley into the city by themselves.

The Lilleheis believed in measured discipline; the consequence for their children's bad behavior was an examination of the issue, not the back of the hand. One day when Walt was coming home from school, a classmate jumped onto the running board of his Model T, which Walt had bought with his savings from caddying. Walt wanted the classmate to get off. When the classmate refused, Walt put the gas to the floor and tore off, fishtailing down the road—and flipped his car rounding a bend. No one was badly hurt, the car still ran—and Walt got away with a lecture from Dad.

"I didn't necessarily believe in signs that said 'don't do this' or 'don't do that,' " Lillehei said years later. "If I had a reason to do it, I usually did it."

A better clue that Lillehei was destined for something other than middle-class anonymity was his skill with his hands. Lillehei instinctively understood how things worked—and how he might make them work. As an eighth grader, Walt successfully modified a BB gun to shoot .22-caliber bullets. As a young teen, he begged his parents for a motorcycle; they resisted. When Walt found some motorcycle parts for sale—a tangled mess in a couple of bushel baskets—his parents, thinking he'd never assemble them, let him buy them. Without benefit of a manual, Walt built the motorcycle and got it running. And when he bought his Model T, Walt slung a hoist over a tree limb, took the engine out, broke it down, and reassembled it with ease.

Although he skipped two grammar school grades, Lillehei became an average high school student who nearly flunked chemistry. He won't last six weeks in college, the chemistry teacher said to Walt's father on Walt's graduation day, in 1935. That fall, at the age of sixteen, Lillehei entered the University of Minnesota. He thought he'd become a lawyer, an engineer, or possibly a dentist, like Dad, but when he learned the requirements for medical school were the same as for dentistry, he figured: Why not? With the exception of three C's, including one for surgery, Lillehei's grades at the University of Minnesota Medical School were outstanding. He graduated tenth in a class of 103.

But Lillehei was no bookworm; he had a wry sense of humor, he loved to carouse, and, after graduating from high school without having dated a girl, he developed an eye for the ladies. He coined his own motto, which he carried throughout his life: Work hard, play hard! Consciously or not, he already was a kindred spirit to John Hunter, the renowned eighteenth-century Scottish surgeon whose biography Lillehei had savored. Hunter was a tireless experimenter who was the first to demonstrate that surgery could be more than glorified butchery—that a surgeon could also be a scientist. But Hunter was no scholarly straight arrow; as a young man, he was drawn to London's gin dens and bordellos, and his life-long disregard for risk bordered on recklessness. In an effort to demonstrate that gonorrhea and syphilis were manifestations of the same disease, Hunter, while engaged to the woman he would marry, infected himself with a pus-laden lancet. "This was on a Friday," observed Hunter. "On the Sunday following there was a teasing itching in those parts, which lasted until the Tuesday following."

During Lillehei's college years, he and his friends spent Saturday afternoons at University of Minnesota football games.

On Saturday nights, they drove out to Mitch's, an establishment on the edge of town run by a one-time bootlegger who booked Hoagy Carmichael, Jack Teagarden, and other Dixieland jazz greats. Mitch's was a bottle club: you could drink all you wanted, provided you came with your own. Walt and the boys arrived with bottles of near-beer, which they brought up to strength with grain alcohol pilfered from a medical school lab.

Many nights, the jamming continued until dawn. Lillehei's stamina was unsurpassed. With only an hour or two of sleep, he was ready to take on the day, no matter what he'd had to drink.

———

When he was a boy, Owen H. Wangensteen did not intend to become a doctor. He wanted to be a farmer, like his father, a Norwegian immigrant who also ran a prosperous general store in the prairie town of Lake Park, Minnesota. Owen would likely have excelled: when he was a high school junior, he devised an ingenious new way to deliver some three hundred piglets from sows that, unable to farrow their young, had been destined for the slaughterhouse. Owen intended to study veterinary science at the University of Minnesota, but his father, a stern man who punished his young children by paddling them out in the woodshed, insisted on medicine. The son finally acceded to the father's wish during the unbearably hot summer of 1917, when Mr. Wangensteen made Owen haul manure until his university classes resumed. "I said, 'Well, anything would be better than this,'" Owen later recalled.

Gifted with a photographic memory, Wangensteen was first in his class at Lake Park High School and the University of Minnesota Medical School. After completing his surgical residency and earning a doctoral degree (for study of the undescended testis), he declined an offer to go into private practice at an annual salary of $15,000, a princely sum. He wanted to teach as well as to operate, and he wanted a lab. So he became an instructor in surgery at the university—for less than seventy dollars a week. Owen's father was disappointed, as was Owen's new wife, a woman who'd majored in home economics and believed a doctor husband was a ticket to the good life. But Wangensteen stood firm.

It was 1926, and the University of Minnesota Medical School was a pale imitation of the famous medical schools back east. European professors rarely stopped by on their lecture tours of America, nor did famous researchers covet a university appointment. The school did not even have a full-time chief of surgery. No one wanted the job.

"Well, there isn't anything here, nor will there ever be," said a Harvard surgeon who declined the position after visiting Minneapolis.

Owen Wangensteen thought this was balderdash. He sensed a rare opportunity, and, after a year of study in Bern with one of Europe's preeminent surgeons, he accepted the position of associate professor at Minnesota. The next year, he was named chief of surgery. Wangensteen was thirty-one years old—perhaps the youngest chief of surgery ever at any university, but already a formidable clinician and researcher, with nearly three dozen published papers on such disorders as bowel obstruction and ulcers, two of his favorite subjects.

Wangensteen had barely taken office when grumbling opponents accused him of losing too many cancer patients. What his opponents really wanted was less an improved survival rate than for the chief to resign—they were jealous, and Wangensteen knew it. "This is a bunch of trivial drivel," Wangensteen told the dean, who kept him as chief.

Politically nimbler now, the new chief began to build his department.

———

Lillehei was a sophomore in medical school when he first saw Wangensteen. It was the fall of 1938, and Lillehei was just twenty. Wangensteen was lecturing.

25

Short and slight, Wangensteen nonetheless was impressive: with his white lab coat, wire-rimmed glasses, and slicked-back hair, he resembled one of the great European professors. And he was young—not yet forty, yet he had been chief for nearly a decade.

Listening to him, Lillehei understood why. Wangensteen expounded for two hours without notes. He brought in patients and presented the case history of each entirely from memory.

The subject was acute appendicitis—the bursting of the appendix—which can kill without surgical intervention. Wangensteen told his students that only a short while ago, the cause was mysterious. Was it infection, as many believed? Skeptical, Wangensteen set out to find the answer.

He initially studied animals. With his research assistant, surgeon Clarence Dennis, Wangensteen observed, measured, and monitored internal pressures after tying animal appendices off. Wangensteen knew nature sometimes hides her secrets in unlikely places, and so he became familiar with the innards of rabbits, dogs, cats, and skunks. Then he moved on to monkeys and apes.

Still, he wasn't satisfied. The young Lillehei listened spellbound as Wangensteen described experimenting on a tiger and a bear, which had been subdued with raw meat laced with barbiturates.

"When they became groggy," Wangensteen recalled, "we rushed up with an ether can and put it over the animal's nose and carried him, with the help of four hands, to the operating table—and then went ahead with surgery after tying the animal down."

Having exhausted the available animals, Wangensteen studied humans. During surgery on colon cancer patients, he performed a second operation that left the patient's appendix sticking out of his abdomen like a misplaced finger; then he tied off the tip and connected it to a bedside meter. And what

Wangensteen found was that obstruction, creating internal pressure, caused an appendix to burst, and not infection. It was no earth-shattering discovery and it had limited clinical value, but it proved the merit of Wangensteen's philosophy. It proved what determination and original thinking could do.

Amazing, thought Lillehei. Wangensteen had embarked on a quest. Here was science in two dramatic environments: the operating room and the lab. Here was Hunterian medicine.

Geez, Lillehei said to F. John Lewis, his closest medical school friend. We ought to go into surgery!

———

Lillehei finished his internship in the spring of 1942, and war summoned him. He went enthusiastically. He believed in serving his country, and he wanted to see the world.

He first saw London, then northern Africa, where, with the 1st Infantry Division, he commanded a mobile army surgical hospital (MASH) unit in the campaign against German field marshal Erwin Rommel's infamous Afrika Korps. Lillehei was introduced to the brutal realities of war even before the enemy was engaged: on the eve of combat, Lillehei and his staff were up all night treating American soldiers who'd shot themselves in the foot, a ploy to escape combat. Lillehei was dumbfounded that anyone could be so scared. He himself was fascinated by war, as John Hunter had been two centuries before. He was heeding the advice of Hippocrates, who declared: "He who wishes to be a surgeon should go to war."

From Africa, the Allies crossed the Mediterranean to Italy, where Lillehei participated in the landing at Anzio, thirty miles south of Rome. It was early in 1944. Anzio was crucial to the Germans, and they defended it fiercely, using punishing air and ground attacks to create what an American general called a "flat and barren little strip of hell." Red crosses on tent tops failed to protect, and U.S. casualties mounted; for days on end, Lillehei did not sleep.

But this was the very heart of war, and the twenty-five-year-old captain found lessons in the carnage—lessons beyond the more obvious ones of management and medicine.

In one letter home, Lillehei wrote: "I've certainly seen more of the horrors of modern warfare than I had ever anticipated . . . you have to see it to believe it. After being around here, it certainly is going to seem funny to go to a football game or something for excitement." In another letter, the young captain observed: "As commander of a hospital, one of my duties is to pick those who have been severely wounded due to direct enemy action and award them the Purple Heart. It is a beautiful medal, but not much to give a man in return for his arm, leg or face." Years later, Lillehei would hope to reward sacrifice with something far greater.

As German resistance weakened and the Allies advanced to Rome, Lillehei found opportunity for other pursuits. Having kept detailed records of all of the operations that his surgeons performed, Lillehei began an analysis of the treatment of war wounds—which he hoped someday to publish. He frequented Rome's cafes with fellow officers and pretty nurses—and he sent love letters home to his girlfriend, the beautiful Kaye Lindberg, a student nurse whom he'd met while he was an intern in Minneapolis and whom he would marry a year after returning from war. Lillehei savored a hearty meal and a night of good drinking, but he mailed most of his officer's pay home to his father, who invested it for him in a savings account. Years later, this would turn out to be fortuitous for Walt, for a reason he could never have imagined.

Lillehei was a lieutenant colonel when he returned to Minnesota, and he wore a Bronze Star, a Bronze Arrow Head, and a European Theater Ribbon with five battle stars. If he'd been even-mannered going to war, he was unflappable coming home. He was a twenty-seven-year-old without discernible temper or fear—a man who was now strangely driven.

"The one change that has occurred in myself since leaving

home is that I've become intensely restless," Lillehei wrote before leaving Italy. "If I've stayed longer than 4–5 days in any one place, I long to get on the move again."

———

Lillehei sailed home late in 1945. Primarily an administrator during the war, he still intended to become a practicing surgeon. It was time for his residency.

He applied to only one program. You do not forget a man who operates on tigers and bears.

When can you start? Wangensteen said at the end of a brief interview.

Today, Lillehei said.

You'll need a white coat, the chief said.

———

Although it owed a debt to John Hunter, Owen Wangensteen's surgical training program was unusual, if not unique. Unlike physicians at some of the European and eastern American centers and even the world-famous Mayo Clinic, ninety miles south in Rochester, Minnesota, Wangensteen was obsessed with research. Many of the great surgical centers emphasized the operating room, where blade met living tissue, but Wangensteen thought that was the easier half. What a surgeon did in the lab—whether he proved to be an original thinker—was more precious, in his view.

"Tradition is good for the French Foreign Legion or the Cold Stream Guards," Wangensteen liked to say, "but it's a disaster in science." (Residents had a saying of their own. "Nothing is too strange as far as Wangensteen is concerned," they said, though never to his face.)

Despite the intellectual freedom he allowed his residents, Wangensteen tightly structured his training program. He demanded regular publication and membership in professional associations. He insisted on master's degrees for all of his surgeons and he encouraged the brightest to earn a doctoral, as he himself had. C. Walton Lillehei was unquestionably one of the brightest.

Lillehei began his residency on January 1, 1946, but before sending the young man to the lab, Wangensteen assigned him to the operating room, for an education in basic surgery. Lillehei started like everyone, by holding retractors and tying knots. He worked the University Hospital wards, ordering X rays, removing stitches, and changing dressings. Ward duty was essential to his training but dull—scut work, they called it. Yet here on the wards was where Lillehei learned to relate to patients, many of whom were terrified of the knife. For a surgeon, even a young one, Dr. Lillehei was unusually compassionate. And that was another lesson he brought home from war: the importance of comforting the sick and dying— a doctor's moral obligation, Lillehei later called it.

Lillehei spent twenty-one months on this beginner's rotation, and it was then—as he was gaining confidence in operating on the intestines, the stomach, the liver, and the lungs—that he had his first flirtations with the heart.

For that, Lillehei owed Richard Varco—like Clarence Dennis, a senior member of Wangensteen's staff. Following groundbreaking work in Boston, Baltimore, and Stockholm, in the 1940s Varco was operating on peripheral structures of the heart. The interior remained beyond reach, but surgery on the outside anatomy was nonetheless a remarkable accomplishment. How amazed the world had been in 1938, when Harvard's Robert E. Gross had ushered in the era of closed heart surgery—surgery that did not require cutting the organ open—with his daring operation on a girl born with a defect of the pulmonary artery and the aorta, the main vessel coming off the heart. How electrifying was the news six years later, when Alfred Blalock of Johns Hopkins University performed the first blue-baby operation, the lifesaving remedy (but not cure) of another deadly congenital heart defect!

Although Wangensteen had performed the first closed heart

operation at the University of Minnesota, in 1939, he found cancer and ulcers more to his liking. And so the chief had handed the closed heart surgery to Varco, a gifted teacher who in turn had initiated Dennis.

Observing Varco operate on the heart's exterior, Walt Lillehei began to get ideas.

———

By the autumn of 1947, Lillehei was itching for the lab. Wangensteen assigned the young resident to his.

Wangensteen's laboratory percolated with new ideas: since becoming chief, Wangensteen, assisted by a succession of young doctors, had created many surgical innovations. Perhaps none benefited medicine as much as the gastric suction tube, used to treat bowel obstruction, a naturally occurring ailment that often killed the patient. Experimenting in his lab, Wangensteen had found that a simple flexible tube passed down the nose and through the stomach to the intestines relieved the pressure of backed-up gas and fluids—the actual cause of death. Used also to help treat abdominal wounds, the Wangensteen Tube had made its inventor something of a hero during World War Two, when so many soldiers took lead in the belly—and later prompted humorist Ogden Nash to pen an ode to the chief:

> *May I find my final rest in*
> *Owen Wangensteen's intestine,*
> *knowing that his masterly suction*
> *will assure my resurrection.*

Wangensteen's obsession in the autumn of 1947 was the peptic ulcer, in which gastric juices eat through the stomach wall. Ulcer was not the most glamorous disease around which a scientist could build a crusade, but a hole in the stomach could kill; even when it did not, an ulcer could make life painful and bloody. Surgeons and nonsurgeons—the in-

31

ternists—often disagreed over treatment. Ulcer was a modern field in the centuries-old contest between the men with knives and the lettered men of medicine, who hated bloodying their hands.

Led by such doctors as the University of Minnesota's Chief of Medicine Cecil J. Watson, the internists preferred to treat ulcers with baking soda or a diet rich in milk and cream; surgeons such as Wangensteen, of course, wanted to cut. The truth was no one had found a universal cure, for no one understood what caused an ulcer.

Wangensteen and his young researchers sought answers in the usual subjects, laboratory dogs; they sought to determine precisely which part of the stomach offended, and exactly how big a piece a surgeon should then take. Lillehei plunged right in. Such adventure—right up there with tigers and bears.

Lillehei spent a year in Wangensteen's lab, and then the chief handed him over to Maurice B. Visscher, chairman of the medical school's physiology department. Like John Hunter, Wangensteen believed no surgeon could excel without knowledge of physiology, the study of function in living organisms.

Lillehei spent a year in Visscher's lab and then, in October of 1949, Wangensteen appointed him the chief resident. Lillehei returned to the operating room, where he would finish his long training—and then join the faculty.

By now, surgeons in Boston and Philadelphia had taken another step toward true open heart surgery: they had devised a way to repair mitral stenosis, a comparatively simple defect inside the heart that involves a narrowing of the mitral valve. Like Alfred Blalock's blue-baby procedure, the mitral valve operation had sent excitement through the surgical world.

Senior surgeon Varco had learned this latest cardiac surgery, and now he taught it and all of the other closed heart operations to Lillehei. Walt had yet to decide his life's work, but the pull of the heart was strengthening.

32

KAYE LINDBERG LILLEHEI

3

Invasive Procedures

WHEN WALT LILLEHEI found a small bump just in front of his left ear while he was shaving one morning in late 1949, he figured it was harmless. Sooner or later, it could be removed, but Lillehei was in no hurry. He was the chief of surgery's handpicked chief resident, he was busy learning closed heart surgery, and already he didn't have enough hours in the day.

Lillehei finally found time to have the bump removed on February 9, 1950. Surgeon David State operated. State specialized in the parotid gland, which lies dangerously close to the nerve that controls facial movement. Careless doctors sometimes cut that nerve, causing a patient's eye to droop or smile to permanently disappear, but State easily excised the bump—not even a brush with danger.

Skeptical of the University Hospital pathology department's initial diagnosis of lymphosarcoma, Wangensteen asked for another study; this time, all of the department's pathologists unanimously concluded that Lillehei had lymphosarcoma. Still unconvinced, Wangensteen described the case to a pathologist friend at the Sloan-Kettering Institute,

America's leading cancer center. The friend was less gloomy than his Minnesota colleagues, but he needed a specimen for his own study to be absolutely sure of the diagnosis. Wangensteen sent him another piece of his prize pupil.

"You remember the suspicions I expressed when we were talking about the parotid lesion," the friend wrote back. "I am sorry, but in this case I would have to admit that the pattern was that of lymphosarcoma."

Wangensteen sent yet another specimen to a pathologist at Columbia University, who not only confirmed the diagnosis but said the cancer was spreading. The cure rate, wrote this pathologist, was 25 percent at five years, "and a somewhat lower one at ten." A fourth specimen sent to the Mayo Clinic elicited a similarly grim response.

It was the middle of March 1950. A month had passed since David State had excised the bump on Lillehei's face.

Wangensteen still did not tell Lillehei—or anyone. He wanted Lillehei to finish his residency. Pathologists' reports notwithstanding, Wangensteen refused to believe.

In April, Wangensteen and several of his surgeons boarded a plane for Denver. From Denver they traveled to Colorado Springs' Broadmoor Hotel for a meeting of the American Surgical Association, whose membership included the biggest names in surgery. During the scientific sessions, new research would be unveiled. Lillehei himself was to present a paper, which would then be published in a leading journal.

This was, in essence, his national debut.

The Broadmoor was an appropriate setting for such an auspicious event. Its interiors had been designed by the same New York firm that had conceived the Ritz-Carlton's Crystal Room, and more than a hundred Italian artists had been hired to paint the ceilings and trim. Guests played polo and golf in the summer, and during the winter, they swam in the indoor pool, filled with pure Rocky Mountain water from the local springs. Only people who mattered made it to the Broadmoor. What a wonderful place for Lillehei and the boys to knock back a few after the last scientific session of the day.

Wangensteen had insisted that Lillehei share his room. Lillehei was honored; the chief wasn't so solicitous of just anyone. What Lillehei didn't know was that Wangensteen wanted to closely observe him. He wanted some symptom—empirical proof that the pathologists were right.

Wangensteen found none, certainly not on April 20, the day Lillehei took the podium.

Few of the surgeons in the audience had ever seen Lillehei, and if they'd heard anything about him, it was only from Wangensteen, who was always crowing about his men. At thirty-one, Lillehei was one of the youngest doctors in the room—and one of the most stylish, in his alligator shoes, gold watch, and three-piece suit. He wasn't nervous. He was remarkably self-assured, even for a surgeon.

He seemed the picture of health.

Lillehei's topic dated from the year before his chief residency—the year he'd spent in physiology chairman Maurice B. Visscher's lab. One of Visscher's interests was the biology of chronic cardiac failure: the complex changes a sick heart precipitates elsewhere in the body, notably the kidneys and the lungs.

Visscher was frustrated in his work by a shortage of experimental subjects. In dogs, the animal of choice for heart research, spontaneous failure was rare. And induced failure was iffy. You could inject a dog's heart with the bacteria that cause endocarditis, and sometimes you got heart failure; sometimes you got a dead dog. You even sometimes got a dog that continued on its merry way. Many scientists had sought a solution—all fruitlessly.

Walt, said Visscher to Lillehei, I want you to devise a reliable method to produce chronic heart failure in dogs.

37

In a matter of months, Lillehei had.

The Broadmoor audience could hardly believe his solution. Why hadn't they thought of it? It was profound in its simplicity! By connecting an artery to a vein in an easy operation, Lillehei had increased the blood flow to the dog's heart, which then had to work harder. After about a month, the stress was overwhelming and the heart began to fail.

In support of his results, Lillehei showed slides of the graphs and charts he had drawn, the X ray he had taken, and photographs he'd made of the autopsied hearts of two subjects, dogs number 24 and number 114; even at the beginning of his career, Lillehei illustrated his findings with powerful images. Lillehei left the stage, then sat down for the rest of the program with his best friend, John Lewis—the two young surgeons wanted to discover all that they could. And one thing they would discover at Broadmoor was the work of an obscure Canadian surgeon who believed that he had devised a way to get safely inside the living human heart.

One day in late May of 1950, Wangensteen called Lillehei into his office. It was a month after Broadmoor. Lillehei was just days from completing his chief residency and moving on to the faculty.

Sit down, Wangensteen said. I have to tell you something.

Lillehei sat.

Remember that node? said Wangensteen. It came back positive for lymphosarcoma. I sent it to three other places, and the diagnosis was the same.

Lillehei was flabbergasted. He'd forgotten all about the bump.

I'm sorry, Walt, Wangensteen continued. We have no choice but to operate.

Wangensteen spoke as if to a stranger. He was not the emotional type, yet Lillehei's cancer hit him hard; here was a man he loved, in his way, like a son. How much of himself Owen

Wangensteen saw in the prize pupil! What intelligence and drive! Nothing got between Walt and his work. The story was told of his needing an X ray on a dog late one night; unable to find anyone with a key to the radiology room, he took the door off its hinges. If anyone could fulfill Wangensteen's fondest desire, a Nobel prize for Minnesota, surely it was Walt.

But there was something else involved—something Lillehei did not, at that time, suspect.

Wangensteen never discussed his personal life, perhaps because it embarrassed and troubled him so. From an early age, Owen's first son and namesake had been a disappointment. Bud, as everyone called the boy, had a genius IQ and a photographic memory, like Dad, but he disliked the rigid structure of school and he defied authority—despite the harsh discipline he received at home for his transgressions. Drawn to a bad crowd, Bud was barely a teenager when he was arrested for stealing cars. Handsome and tall with a charm that gave no clue of his volatile temper, Bud eloped when he was eighteen, became a father when he was nineteen and again at twenty, dropped out of college twice, and, when Owen stopped giving him money, sold vacuum cleaners and encyclopedias door to door—and forged stolen checks, a crime for which he was imprisoned. Bud gambled and drank, and fancied himself a writer, a young Hemingway.

Wangensteen might have found comfort in his wife, but his marriage had proved a disaster. Pleasant-looking and popular, Helen Carol Griffin was a city girl from St. Paul. She was enrolled in the University of Minnesota's College of Agriculture, Forestry and Home Economics when she met Owen, a shy farmer's boy. Helen married Owen and gave birth to their first child, a girl, while her husband was still studying as a resident. But a doctor would surely give her the privileged life for which she yearned! How disappointed was Helen, then, when Owen took a university job at a fraction of his potential salary; how her resentment grew seeing her sister, wife of a wealthy radiologist. Owen drove an old car, wore old suits, and carried the same old briefcase; the only money he desired was for his department's budget.

Compounding things was Helen's increasingly heavy drinking, her bouts of depression, her invariable siding with Bud in Bud's many entanglements with his father. Helen and Owen Wangensteen had come to hate each other—and Helen's hatred was cruel. Knowing how her husband feared serpents, she wrapped a dead snake and gave it to him one year as a birthday present. Another time, she substituted a nasty personal letter for a speech he was to deliver to an academic audience. Owen did not discover this until, at the podium, he reached into his pocket and pulled out Helen's ranting.

Now Walt's lymphosarcoma. How unfair life could be.

———

It was May 25, 1950, the start of the Memorial Day weekend. Lillehei went home to assess his options.

He had only a few. There was no chemotherapy in 1950, no bone marrow transplantation. Radiation without surgery was a possibility, but Wangensteen had advised against it; he believed Lillehei's long-term chances would be better with the knife. Wangensteen offered to operate.

This was less than reassuring.

Wangensteen had more cancer experience than anyone at the University of Minnesota, but Lillehei, like others familiar with his work, knew his aggressive philosophies. Wangensteen only too willingly took on cases that other surgeons had abandoned as hopeless. He believed timidity had no place in this field—that when in doubt, it was better to take more than less, for cancer cells could be lurking anywhere. Among the operations he sanctioned for the worst malignancies were the hemicorporectomy, which basically involved cutting a person in half, discarding both legs in their entirety; and the eviscerectomy, in which the surgeon removed the bladder, the reproductive organs, the lymph nodes, the spleen, the rectum, a kidney, and all but about a foot of the colon. Believing a devoted surgeon could do more for a woman with severe breast cancer than perform a traditional radical mastectomy, Wangensteen advocated the so-called super-radical mastectomy, which involved the deepest possible cutting.

Wangensteen's critics said his audacity clouded his judgment. "Wangensteen hates cancer because it kills more people than he does," some of his own residents used to say. And after word leaked out of University Hospital about a hemicorporectomy one of Wangensteen's surgeons had performed, the surgeon received a letter. "Now that you've done that," the letter said, "why don't you cut off your own head."

But Wangensteen never lost sleep over criticism. His newest passion, already controversial, was the so-called second look. Starting with the widely agreed assumption that a surgeon could never be sure he'd gotten every cell of metastatic cancer, Wangensteen recommended a second operation half a year later, even if the patient had no symptoms of recurrence. A patient could look and feel fine and Wangensteen still wanted to open him back up. If traces of cancer were found, a second look could lead to a third or fourth—or more. In one of his first articles on the subject, Wangensteen profiled a sixty-year-old woman who, the chief wrote, had displayed "a remarkable degree of courage, faith and patience." Others might have called her pathetic. Operated on for cancer of the bowel in 1948, she returned to Wangensteen's operating room not once but five more times. "This sixth look," wrote Wangensteen, "was a careful general exploration." Fortunately for the woman, this time she was cancer-free. Wangensteen undoubtedly would have summoned her again.

———

After reading what he could find about the treatment of lymphosarcoma, Lillehei reluctantly agreed to surgery. Trying to stay focused on his work until the very last minute, he was awake into the early hours of the day of his operation completing the final draft of an article about ulcers that he'd coauthored with Wangensteen. Before finally going to bed, Lillehei asked his wife, Kaye, to type it.

At 7:15 A.M. on Thursday, June 1, Lillehei entered University Hospital's Room I, Wangensteen's room. He did not know exactly what would be done to him while he was asleep. He knew only that Wangensteen intended to open him up and remove everything that might conceivably be cancerous.

David State, the surgeon who'd removed Lillehei's parotid tumor in February, began the operation.

Under Wangensteen's supervision, State excised the remainder of Lillehei's parotid gland. Then senior surgeon Varco scrubbed in and he and State started on Lillehei's neck, from which they took all of the lymph nodes and glands. Some of the nodes near the jugular vein were enlarged, and Wangensteen decided they had to go down into the chest. Wangensteen had not raised this possibility to his patient, but it was too late now to seek permission: Lillehei was dead to the world, his face and neck splayed open like an anatomy-class cadaver.

Now Wangensteen scrubbed in. Four hours had passed; for 1950, it was already a marathon.

Assisted by yet another surgeon, John Lewis, Lillehei's best friend, Wangensteen split the sternum and opened the chest. Wangensteen carved deep, removing more lymph nodes, more glands, muscle, fat, vessels, the thymus, an entire rib. No operation like this had ever been done anywhere. This was scorched earth, and the bleeding was horrendous. Lillehei was transfused, pint after pint after pint of blood—including one donated by Norman E. Shumway, an intern who many years later would invent human heart transplantation.

Ten hours and thirty-five minutes after the operation started, Wangensteen was finally done.

Seven surgeons, four anesthesiologists, and several nurses had assisted.

Nine pints of blood had been used.

Twenty-three specimens had been sent to the pathologist.

Lillehei faced twelve sessions of radiation.
Still, he was alive.
The odds said that in five years, he would not be.

A week after Wangensteen's lymphosarcoma surgery, Lillehei was discharged to the care of his wife. They lived with their young daughter in a duplex apartment near the university.

The only daughter of Swedish immigrants, Katherine Ruth Lindberg grew up in Minneapolis. She was an uncommonly pretty girl, who was voted Most Popular by her high school classmates—and who had her choice of boys. "If you had listened closely, you would have heard my knees rattle," one of her high school suitors confided in a note he slipped to Kaye after watching her play volleyball, one of several sports at which she excelled. "You were the cutest one on the floor— thanks for the privilege of looking at you. I think you are perfectly proportioned. . . . What is your locker number?"

After graduating at the top of her class, Kaye entered the University of Minnesota's nursing school, intending to become a stewardess and then a practicing nurse, perhaps even a supervisor or administrator. Kaye met Walt in 1941, at Minneapolis General Hospital, where she was studying and Walt was serving his internship.

Wow—look at that blond! a friend of Kaye's said to her one day when Dr. Lillehei walked onto the ward.

Kaye agreed that Walt was a looker. And she admired the way he, unlike so many of the interns, always took the time to listen to patients and offer them encouraging words. For his part, Walt thought Kaye had the best legs of any of the student nurses.

Kaye was dating someone else, but when he left for the navy, Walt asked her to a hospital picnic. They went steady from that day until Lillehei enlisted in the army, in June of 1942. Before leaving, Walt gave Kaye his fraternity pin; when the war was over, they would marry. For more than three years, as Lillehei moved with the Allies across northern Africa and into Italy, the couple exchanged letters constantly—and planned to reunite overseas even as war raged.

"My darling Kaye, Sweetheart," wrote Walt in one of his letters, "I'm so damn much in love with you I'm in misery. . . . You are so darn cute and lovely darling that you undoubtedly will get many invitations for dates, but please wait for me faithfully my dear because I am sure that we will be together very soon."

But Kaye never did get overseas during the war. And three and a half years apart took a toll: although Kaye and Walt remained engaged after Walt came home, they did not rush to the altar. Wangensteen's demanding residency program absorbed Walt, and Kaye was flying for Northwest Airlines. "We were two different people," Kaye recalled later. "We just sort of went on different paths."

An abrupt change in airline policy pushed Kaye and Walt to wed. Desiring only unmarried stewardesses, Northwest in late 1946 declared that starting in 1947, married women would no longer be hired; the only married stewardesses would be those already married and on the payroll by that January 1. Figuring it was now or never, Kaye married Walt on New Year's Eve—beating the deadline by mere hours. Eighteen months later, the Lilleheis had their first child, a girl they named Kim.

———

Never before gloomy, Lillehei went into a funk the summer of 1950. His chest wound became painfully infected, and Varco came by evenings to clean it out (for his troubles, Lillehei mixed them both martinis). Another complication, a dilated stomach, sent Lillehei staggering to the emergency room. Two weeks of radiation treatment for his face left him nauseated and raised the specter of worse side effects some day, including cataracts. That, of course, was assuming Lillehei lived.

This was no easy time for Kaye, either. Doctors had just sent her mother to a sanatorium for tuberculosis, and with the care of her sick husband, now Kaye had to abandon her work toward a master's degree in nursing, which she had been pursuing at the University of Minnesota. Her stewardess days were already history; abruptly reversing its marital-status policy, Northwest Airlines had sent the newlywed a pink slip shortly after she'd returned from her honeymoon.

That terrible summer, Walt tried to move his mind off things by watching TV, mostly afternoon baseball games and a few programs that were broadcast before midnight, when stations signed off until morning. He read medical journals, worked some on his doctoral dissertation, and looked ahead to autumn, when he hoped to resume operating and open his own lab.

Physical suffering was only a part of that summer's misery. Although he never mentioned it to Wangensteen, Lillehei resented the chief's cutting so deep, without forewarning—especially considering the pathologist's final report, which showed no further malignancy anywhere. Nonetheless, Wangensteen had recommended a second-look operation in six months. Lillehei refused. Enough was enough.

And Lillehei worried for his young family. Wangensteen continued to pay his salary, but no insurer would cover a lymphosarcoma survivor. With savings from the wartime pay he had dutifully sent home, Lillehei began to invest in the stock market. This, too, would turn out to be fortuitous in ways the young doctor could never have imagined.

DOROTHY EUSTICE

4

Blue Babies

LILLEHEI RETURNED to University Hospital after four months of convalescence. His first operation, by coincidence, was excision of a parotid gland; he performed it easily, although he could never split a patient's sternum without remembering his own painful operation and lingering infection.

A junior member now of the University of Minnesota Medical School faculty, Lillehei opened a small lab in the attic of the physiology building—where, in a larger lab across the hall, senior surgeon Clarence Dennis was refining his heart-lung machine, a project Lillehei followed with keen interest. Lillehei's first independent research was an extension of his study of chronic heart failure in dogs—a study that had brought his first national award: a medal and $1,250 from the American Association for the Advancement of Science. With the money, Lillehei began to investigate the effects of shutting off blood to the hearts of dogs, determining how long they could last without brain damage. In the operating room, Lillehei resumed closed heart surgery and general surgery, which provided steady income.

It was the autumn of 1950, the beginning of an unusually prosperous decade. Nurseries overflowed with the baby boom and a new drug, penicillin, was keeping babies healthy. Parents were transforming pediatrician Dr. Benjamin Spock into a kindly national hero, and noted cardiologist Dr. Paul Dudley White was preaching the gospel of cardiac fitness—fundamental to the good life. The heart had always been glamorous, and now it seemed an exciting new frontier.

Newspapers brimmed with the latest cardiac adventures— of Soviet researchers claiming to have transplanted frog hearts, of the American ex-Marine who revealed that he'd lived with a bullet in his heart since Okinawa. Even that staid paper of record *The New York Times* could not resist. HEART-SHOCKED BOY LIVES was the headline on a *Times* piece about a child whose heart was jolted back to rhythm by surgeons wielding an extension cord plugged into ordinary household current. Another *Times* story titillated with the promise of an "atomic cocktail" to rejuvenate ailing hearts.

The most dramatic accounts featured people revived after having been declared clinically dead, and the most amazing of those was a patient identified only as the Miracle Man, sixty-five years old and from Long Island. The Miracle Man made the *Times*'s front page in April 1950. The miracle began with an abdominal operation that had turned disastrous.

"At about 1:30 P.M.," the *Times* reported, "the anesthetist reported to the surgeon that the patient was dead, his heart and respiration having ceased. The doctor immediately made an incision over the heart, reached into the chest cavity, and began to massage the lifeless organ by hand." For more than six hours, the surgeon hung on. Eventually, the Miracle Man's heart took over—and he soon recovered and went home. "I expect to get in some fishing with him out on the island this summer," the surgeon said.

Breathless accounts notwithstanding, the state of the art was primitive, and far more heart patients were lost than rescued in 1950. There were no replacement valves in those days;

no pacemakers or defibrillators to rescue a failing heart; no cardiopulmonary resuscitation (CPR) to save a life during a heart attack; no coronary bypass surgery to mitigate the effects of smoking, stress, and fat-rich diets; few cardiac medications; and not even any foolproof diagnostic tests.

There was no way to correct the worst defects—still no way to operate inside the open heart.

———

As the one-year anniversary of his operation for lymphosarcoma approached, Lillehei seemed healthy. He and Kaye were expecting their second child, and to celebrate their all-around good fortune, Walt dipped once more into his wartime savings to buy his first fancy car: a 1951 Buick Roadmaster convertible, with V-8 engine and Dynaflow automatic transmission. "It makes life richer," the ads proclaimed.

In March 1951, at Wangensteen's urging, Lillehei, Kaye, and their daughter, Kim, left Minneapolis in their new car on a working vacation for Walt. He would visit prominent surgeons, heart surgeons especially, in hope of seeing firsthand the latest techniques, some of which had been revealed the previous year at the Hotel Broadmoor.

Lillehei first traveled south: to Iowa, Kansas, Missouri, and New Orleans, places that for the most part occupied the rear guard of cardiac research. The East Coast was another matter. With their esteemed medical schools and world-class teaching hospitals, Baltimore, Boston, Philadelphia, and New York were generally considered the finest medical centers in America—and of these, Boston was first among equals.

In Boston, Lillehei kept an appointment with the distinguished Harvard professor Edward D. Churchill, chief of surgery at the Massachusetts General Hospital; in Churchill's lab nearly two decades before, John Gibbon had begun his groundbreaking work on a heart-lung machine. Then Lillehei crossed town to Harvard's Peter Bent Brigham Hospital, where Dwight E. Harken was chief thoracic surgeon.

Harken was already a legend in closed heart surgery: during wartime service in the army, he had finally, fully, deflated a long-standing notion. From ancient times, philosophers and scientists romanticized and revered the heart, even as it remained mysterious. The heart was deemed a life force—the very seat of the soul, many believed—a part of the body that guarded its secrets ferociously, for no one could so much as glimpse inside of a human heart while the person lived. Experience suggested that, despite its vital importance, the heart was singularly fragile; unlike many lesser organs, it seemed unable to heal itself. "The heart alone of all viscera cannot withstand serious injury," wrote Aristotle, expressing a conviction that went unchallenged for two thousand years. Nor could any doctor accomplish what the body on its own seemingly could not, the ancient Greeks and Romans believed. "Although Aesculapius himself applies the sacred herbs," declared Ovid, "by no means can he cure a wound of the heart."

This philosophy persisted into modern times. Well past the seventeenth century, when the great English physician and scientist William Harvey first accurately described the circulatory system, surgeons believed that the one organ that would forever elude them was the heart. Still, by the early nineteenth century, as medicine in general blossomed, doctors were beginning to suspect that the heart might be more durable than previously imagined. They found that passing a catheter into the chest for drainage of blood and fluids could help a wounded heart heal, and they also prescribed leeches and quiet bed rest—occasionally with positive results. By the middle of the century, several doctors reported successfully treating patients with injured hearts, although overall mortality was high: 90 percent, according to the most publicized study.

Encouraged by this modest success, surgeons contemplated their own interventions, and in 1882 a German doctor (named Block) revealed that he had made puncture wounds in the walls of rabbit hearts, then sewed the wounds up—and the

50

Blue Babies

rabbits had lived. But the suggestion that success in rabbits foreshadowed success in people was met with swift condemnation; ancient prohibitions endured. Said the distinguished Viennese professor and surgeon Theodor Billroth, in 1883: "A surgeon who tries to suture a heart wound deserves to lose the esteem of his colleagues." Thirteen years later, noted British chest surgeon Stephen Paget spoke even more strongly: "Surgery of the heart has probably reached the limits set by Nature to all surgery; no new method and no new discovery can overcome the natural difficulties that attend a wound of the heart. It is true that 'heart suture' has been vaguely proposed as a possible procedure, and has been done on animals, but I cannot find that it has ever been attempted in practice."

In fact, it had: the year before, Norwegian surgeon Ansel Cappelen had sewn a cut to the heart wall of a twenty-four-year-old man who had been stabbed. But the man died a few days after the operation, as did the world's second such patient, a thirty-year-old victim of a dagger fight that Italian surgeon Guido Farina operated on in March 1896. Then, that September 9, German surgeon Ludwig Rehn succeeded. He split the chest of a dying twenty-year-old stab victim; opened the pericardium, the sac that encases the heart; and identified the source of the man's distress, a small laceration of the heart wall. "The wound was closed with three silk stitches," Rehn wrote. "The pulse immediately improved." Eventually, the man recovered and went home.

A barrier had fallen—but many more remained. Well into the twentieth century, the inside of the living heart remained forbidden territory to all but the pathologists.

When Dwight Harken, a redhead with a fiery temper, began his life's work, only a handful of surgeons had ever attempted removal of foreign bodies (needles, nails, shrapnel, and the like) from near or inside the heart. None had claimed great success; such operations, even military surgeons in 1944 be-

51

lieved, should be undertaken only as a last resort, for a corpse was nearly always the outcome.

But Harken was an iconoclast; as a young army surgeon, he questioned such prevailing wisdom. He saw no reason why the heart deserved sacred status, and so, against the advice of his superiors, he decided to put his beliefs to the test. During the D day landing in France, in June of 1944, medics brought a badly wounded soldier to Harken's operating table. An X ray suggested that shrapnel had lodged in the heart's outer wall, but when Harken opened the dying soldier's chest, he discovered that the metal was actually inside the right ventricle, one of the heart's two pumping chambers.

Somehow, Harken managed to insert a clamp through the wound and grasp the fragment without killing the soldier.

"For a moment, I stood with my clamp on the fragment that was inside the heart, and the heart was not bleeding," said Harken in a letter to his wife. "Then, suddenly, with a pop as if a champagne cork had been drawn, the fragment jumped out of the ventricle, forced by the pressure within the chamber. . . . Blood poured out in a torrent!"

Now what?

Harken had placed sutures along the edges of the wound, but tightening them did not stop the bleeding.

Now what?

"I told the first and second assistants to cross the sutures and I put my finger over the awful leak," said Harken. "The torrent slowed, stopped, and with my finger *in situ,* I took large needles swedged with silk and began passing them through the heart muscle wall, under my finger, and out the other side. With four of these in, I slowly removed my finger as one after the other was tied. . . . Blood pressure did drop, but the only moment of panic was when we discovered that one suture had gone through the glove on the finger that had stemmed the flood. I was sutured to the wall of the heart! We cut the glove and I got loose. . . ."

The patient recovered—and Harken was emboldened. Through war's end, the young surgeon removed foreign bodies from inside (or near) the hearts of 134 men without a single death—an unprecedented achievement, one his superiors had never believed possible. Whether the innermost reaches of a badly diseased heart could tolerate surgery was still unknown, but the heart wall, Harken had demonstrated once and for all, could be safely, even routinely, penetrated and sutured.

———

From Harken's Peter Bent Brigham Hospital, Lillehei crossed the street to another Harvard affiliate, The Children's Hospital, where the most celebrated cardiac surgeon of all was professor and chief of surgery.

Even in hallowed Boston, no one compared with Robert Gross. His remarkable surgical cure in 1938 of a little girl with patent ductus arteriosus, a birth defect involving the pulmonary artery and the aorta (but not the inside of the heart), had given hope to those who dreamed that someday every heart-crippled child could be healed. Gross's skill with the knife was legendary, and his contributions spanned virtually the entire anatomy. He had inspired many young doctors to pursue cardiac research—including John W. Kirklin, a University of Minnesota graduate and Lillehei contemporary who, in 1951, already was the preeminent closed heart surgeon at the Mayo Clinic. Before coming to the Mayo, Kirklin had earned his medical degree from Harvard Medical School, where, in a lecture hall, he first saw Professor Gross. Decades later, Kirklin still carried the image of Gross on that long-ago day: a dashingly handsome man whose bearing was imperial. "He walked in through the right side and had to walk all the way across the room because the lectern was on the left side," recalled Kirklin, "and in that walk, everybody became a cardiac surgeon."

But Gross did not welcome Lillehei warmly that April of 1951; had Gross not admired Owen Wangensteen, he might not have received Lillehei at all.

Unlike Wangensteen, Gross as he aged felt threatened by the brightest of his young trainees. He became suspicious of strangers—not without reason. Shortly after his 1938 break-through, Gross began laboratory research on another crippling congenital heart condition: coarctation (narrowing) of the aorta. Among the visitors to Gross's lab was the Swedish surgeon Clarence Crafoord. Crafoord observed the technique Gross had developed for surgery on dogs, and then went back to Sweden, where, in October 1944, using Gross's technique, he became the first surgeon to repair coarctation of the aorta in a human. Gross, who did not have his first human success until 1945, forever believed that Crafoord had stolen his ideas, even though Crafoord himself was a formidable innovator.

When Lillehei called on Gross that spring day in 1951, the young visitor at first was told that the great surgeon's lab was closed. Then Gross relented—but allowed Lillehei only a brief tour. Lillehei got only an inkling of Gross's current preoccupation: the atrial well, an ingenious but ultimately impractical method of entry into the open heart.

Gross was one of the few cardiac surgeons in 1951 who was unconvinced that some sort of heart-lung machine would hold the key to correcting the worst defects inside the heart. "We do not believe any of them will have more than a limited usefulness," said Gross of others' approaches.

Gross favored his atrial well: an open-ended, funnel-shaped rubber bag with which he was experimenting on dogs at the time of Lillehei's visit. After opening the animal, Gross sewed the bag onto the heart wall—over the right atrium, one of the organ's two holding chambers. With the well thus secured, Gross cut into the heart. Pressure forced blood to rise several centimeters into the well, but it was tall enough that no blood overflowed. Working with his fingers submerged in a swirling pool of crimson, Gross managed to sew shut a hole inside the beating heart.

Later, after he'd successfully used the atrial well on people, Gross wrote extensively of the technique, maintaining that, with practice, any respectable surgeon could master it.

"Even though the operator cannot see what is being done," he wrote, "the stitches can be placed and tied accurately by 'feel.' " Gross then described the convoluted steps needed to sew a septal hole—a hole in the septum, or inner wall, that divides the heart's chambers:

> With the left index finger, the margin of the septal defect is palpated; the right hand manipulates the 10-inch long holder carrying the needle. The tip of the needle holder is passed through the septal opening and the needle is made to pierce the septal tissue, 3 or 4 mm. back from its edge, guiding this maneuver with the left index finger tip. The needle-holder is then given to the assistant, who carefully maintains the upward traction and impalement of the needle. With his right hand, the operator passes down into the (atrium) a right-angle clamp; under guidance by the left index finger, the right-angle clamp grasps the tip of the needle. The assistant releases the original needle-holder; the operator pulls up the needle with his right-angle clamp.

And that was just for placing one stitch!

Closing a large hole required many stitches, and often the even more challenging placing of a plastic patch. And never mind that throughout the operation, the surgeon worked blindly—while ensuring that this slippery rubber gizmo did not topple over, get punctured, or detach from the heart wall.

The truth was, the atrial well could succeed only with a surgeon of Gross's abilities. Walt Lillehei was such a surgeon, and, after Gross publicized his technique, in 1952, the young Minnesotan not only successfully used it, he improved it. But

Lillehei soon discarded the atrial well, for he believed it was too complicated ever to be widely used. He believed, correctly, that the atrial well would turn out to be nothing but an odd footnote in the history of cardiac surgery.

Before returning home, Lillehei and his family swung through Maryland and Pennsylvania, where the tradition of medical excellence dated to colonial times.

In Baltimore, Lillehei visited Alfred Blalock, the Johns Hopkins University surgeon who had devised a surgical treatment for tetralogy of Fallot, a defect that prevented proper oxygenation of blood and gave a child's skin a distinctive blue cast during the few, sick years he had to live. The blue-baby operation involved joining a branch of the aorta to the pulmonary artery; it did not cure tetralogy of Fallot, which was actually a cluster of abnormalities inside the heart, but it dramatically extended and improved life. Like Gross's 1938 operation, Blalock's surgery had become a sensation in the popular press when it was reported in 1945—all the more so since Blalock's partner was one of the rare female doctors of that era, the cardiologist Helen B. Taussig. Like Gross, Blalock inspired a generation of doctors, including the promising young surgeon Denton A. Cooley, who in 1951 had just completed a four-year residency at Johns Hopkins.

But Lillehei himself had already performed blue-baby operations with Varco back in Minnesota. Blalock was not in the forefront of the quest for open heart surgery, and so, after a few hours in Baltimore, Lillehei traveled to Philadelphia. There, he passed on a chance to visit Jefferson Medical College, where John Gibbon continued to experiment with his heart-lung machine. Lillehei saw no purpose in visiting Gibbon: the University of Minnesota's own Clarence Dennis had built a device based on Gibbon's blueprints—had, in fact, in Lillehei's absence, become the first surgeon ever to use a heart-lung machine on a person when he operated on Patty Anderson.

So Lillehei introduced himself to another Philadelphia surgeon, Charles P. Bailey. A graduate of Hahnemann Medical College—a second-tier medical school, at least to Ivy Leaguers—Bailey was closer in spirit to John Hunter than to Gibbon or any Harvard professor. As Lillehei would also soon prove to be, Bailey was just this side of reckless. He was a notorious renegade.

Convinced by experiments and autopsy studies that he could repair a stuck heart valve with his finger, Bailey in November 1945 had brought a heart-crippled thirty-seven-year-old man to his operating table. No one had ever fixed a faulty heart valve; except for Henry Souttar, an English surgeon early in the century who had been soundly scolded by his peers for such a scandalous act, no one in 1945—not even the daring Dwight Harken—had so much as touched the inside of a living human heart. The closest a surgeon's finger had ever been was the outside of the wall.

Bailey opened the man's chest and exposed the left atrium of the heart. Then he sewed a purse-string suture into a section of the atrium, clamped that section, and cut into the heart wall. Slowly releasing the clamp, Bailey inserted his index finger; by gently tightening the purse string, he was able to prevent blood from leaking. Bailey had intended to maneuver his finger down to the man's stuck mitral valve, the gateway between the left atrium (which receives freshly oxygenated blood from the lungs) and the left ventricle (which then pumps that blood, under enormous pressure, through the aorta into the body). He had intended to then use his fingertip to free the valve—but he never got the chance. Disease had weakened the atrial tissue, and the purse-string stitches let go. Blood gushed out of the heart. Bailey withdrew his finger and reapplied the clamp—but this time, the clamp crushed the tissue. "Massive uncontrollable hemorrhage resulted with immediate fatality," Bailey later recalled.

But Bailey was not deterred; like Lillehei, he did not recoil from death.

For his second try, in June of 1946, Bailey selected a twenty-nine-year-old woman who was experiencing congestive heart failure. This time, instead of his fingertip, Bailey used a metal punch that looked like a pencil—but it didn't work. The patient's blood pressure plummeted, and her skin turned a dangerous blue. Scrambling now, Bailey withdrew the punch and forced his index finger through the valve, and almost immediately, the woman's blood pressure returned to normal and she lost her blueness. But two days after surgery, she died.

By now, the internists at Hahnemann Medical College were alarmed: behind Bailey's back, people had started calling him "the butcher." One of Hahnemann's most powerful internists, the chief of cardiology, forbade Bailey from attempting any more of his crazy mitral valve cases.

"It is my Christian duty not to permit you to perform any more such homicidal operations," he said.

"And it is my Christian duty to perfect this operation," replied Bailey. "Nothing could be worse than what mitral stenosis does to people."

Bailey may have been an unrepentant risk-taker, but he was no fool. He scheduled his third case, in March of 1948, at an out-of-state hospital: Memorial Hospital in Wilmington, Delaware, where he was a consultant. Once again, Bailey succeeded in the operating room—and failed a short while later on the ward. Seeking to prevent dangerous blood clots during recovery, Bailey unintentionally prescribed an overdose of anticoagulants and intravenous fluids; this caused internal bleeding and damaged the thirty-eight-year-old patient's already weakened lungs, and five days after surgery, he died.

Three attempts, three deaths.

And still, Bailey was not deterred.

But now he was a pariah at two hospitals; it seemed likely that if he did not soon succeed, his worsening reputation might keep him from ever trying again anywhere. And so Bailey scheduled two mitral valve operations for the same day, in separate places: one at eight o'clock in the morning at Philadelphia General Hospital, and the second at two o'clock that afternoon—across town, at Episcopal Hospital.

"If the operations were done at different hospitals," Bailey later explained, "the probability was great that news of a mortality during the first operation would not reach the second hospital in time to interrupt the performance of the later procedure."

Shortly after 8 A.M. on June 10, 1948, Bailey opened the chest of a dying thirty-year-old man. Bailey knew almost immediately the surgery was ill-fated: merely touching the wall of the heart nearly stopped its beating. Bailey administered an experimental drug that he thought might help—and the heart stopped altogether. Bailey brought it back with manual massage, but the heart stopped again whenever he withdrew.

Despairing, the patient's primary physician urged Bailey to open the heart and try to fix the mitral valve. Bailey refused. The primary physician insisted. Bailey finally agreed—but only if the primary physician would first pronounce the patient dead. The primary physician did.

Bailey then opened the heart and forced his finger through the valve. But forty minutes later, the heart stopped for good.

Four attempts; four deaths.

Bailey washed the blood from his hands, changed into his suit, and drove to Episcopal Hospital, where his fifth—and possibly last—mitral valve patient awaited him. Claire Ward was a twenty-four-year-old housewife who, as a child, had suffered from rheumatic fever, an infection that can damage the mitral valve. Ward had lived with a heart murmur until she was an adult, when her health suddenly failed; by the time Baily saw her, she could not care for her young child. She could not draw a good breath, and medication did nothing.

Ward knew of Bailey's failures; her doctor had tried to talk her out of an appointment with the maverick surgeon, but she disregarded the advice. Even as Bailey's morning patient cooled in the morgue, Bailey opened Ward's chest, and, with a hooked knife, widened the damaged mitral valve. After widening the valve further with his finger, Bailey closed Ward back up.

Three days later, Ward was up and walking.

And four days after that, Bailey brought her to Chicago, where he presented her at a meeting of the American College of Chest Physicians. Ward then went home—a healthy woman who would have two more children and live thirty-eight more years before dying of heart and lung failure, probably related to the cigarettes that she smoked.

Dwight Harken, meanwhile, was perfecting an operation similar to Bailey's. Many surgeons in the late 1940s would soon learn the new procedures—including Varco, who in turn taught Walt Lillehei.

———

Unlike Professor Gross, Charlie Bailey did not fear strangers. Visiting surgeons were always welcome in his operating room and in his lab, and Lillehei stayed four days, the longest stop of his six-week journey. He assisted Bailey on a mitral valve operation, learned methods Bailey was developing to repair the heart's other valves, studied the heart-lung machine Bailey was building, and watched Bailey experiment with cooling (hypothermia) as a means of peforming true open-heart surgery.

Lillehei had not been dazzled by anything he'd seen in Boston; Gross's time, especially, seemed to have passed. But Charlie Bailey spilled over with new ideas. Lillehei could barely contain his excitement.

"He is unbelievably clever in developing methods for exploration of all four cardiac chambers," wrote Lillehei in a letter to Varco. "This philosophy, of course, shocks all internists and most surgeons into insensibility. But the man is unbelievably correct. I'm sure of that—and I told him so."

Lillehei missed Dennis's operation on Patty Anderson, but he returned to Minneapolis in time to observe Dennis lose his second open heart patient, two-year-old Sheryl Judge, whose body was pumped full of air.

There has to be a better way, thought Lillehei. Something simpler than all those doctors and that confounded machine.

Back in his lab, Lillehei resumed his research into the effects of blocking off the circulation in dogs. The pull of the heart continued to intensify, but it took a young woman for Lillehei to succumb completely.

Her name was Dorothy Eustice.

Born in 1928, Dorothy was the next-to-the-last of the seven children of Ethel and Thomas Eustice. Ethel had inherited a small farm on the Minnesota prairie, but her ne'er-do-well husband had never made a go of it. Thomas drank, and some of his relatives believed his drinking played a role in the mysterious death in 1924 of two-month-old Lucy Eustice— "found dead in bed with a nipple in its mouth," the medical examiner wrote, but family members believed that in his inebriation, Thomas had rolled over and smothered the infant, who'd been sleeping with her parents. In 1932, at the height of the Great Depression, when Dorothy was four, Thomas walked out on his family for good.

Like Lucy, Dorothy seemed destined for a tragic end. As a baby, her lips and fingers frequently turned blue, and, sometimes, her heart pounded so violently that it shook her bed. After diagnosing an "enlarged heart," a doctor prescribed the heart stimulant digitalis and told Ethel to restrict her daughter's exercise, for little Dorothy often gasped for breath.

Dorothy made it through first and second grade, but then bad health forced her to stay home; she could not walk the two blocks to school. After a year, the girl's health returned, lasting this time until she was a young teenager, when she collapsed while walking to a country fair.

It was 1942, the year Ethel died, leaving her sick daughter a fourteen-year-old orphan. Dorothy went to live with her aunt, and, five years later, she moved in with a sister who lived in yet another prairie town, Janesville.

By now, Dorothy was deathly ill; like Patty Anderson, she was exhausted by stairs, and many nights she slept only by sitting up in bed. She suffered from vertigo, red blotches sometimes broke out on her face, and her abdomen was bloated, a sign that her kidneys were failing. Dorothy's pulse would suddenly jump to two hundred—and then her heart, grossly enlarged now, would drop back to a normal pace.

"It just goes fast and then slows down," Dorothy told a doctor on her first admission to Minneapolis's University Hospital, in January of 1948. Doctors diagnosed two deadly problems: mitral valve stenosis and atrial septal defect, or ASD. The doctors made Dorothy comfortable and then sent her home; in early 1948, there was little else they could do.

Dorothy was nineteen, a young woman. She was bedridden much of the time by then, and her sister made over the first-floor parlor of their small house into a bedroom; this saved Dorothy from stairs and kept her near the kitchen, the social center of the family. On bad days, Dorothy took her low-salt meals in bed—but on her better days, she sat in a wooden chair that a relative had built with a special high-placed seat that was easier to get in and out of. The chair was positioned by the window, where Dorothy could watch the children play and admire the flowers that her sister planted. Dorothy did not have a television set, but she listened to the radio and she liked to read. She liked visitors. They brought news of a world that no longer was hers.

———

Lillehei met Dorothy Eustice in November of 1951. Walt's mother had heard about the young woman from a friend, and Mrs. Lillehei insisted that her doctor son stop by Dorothy's room to cheer her up. Dorothy was by then on her sixth admission to University Hospital; twenty-three years old, she'd lived longer than her doctors had ever imagined she would.

Lillehei was moved by Dorothy's attitude—how she seemed at peace, despite being so sick. Dorothy delighted in simple things: crocheting, animals, the vanilla ice cream and Hershey Bars that her sisters snuck her against doctor's orders.

And Lillehei was struck by Dorothy's pale beauty. The woman did not have the defeated look of a chronic invalid; she resembled a china doll, with red hair, porcelain skin, and brown eyes that gently drew you in. Lillehei, of course, knew that she was dying; listening to her heart, he heard a terrible whining, like a string frantically vibrating.

Although surgeons using Charles Bailey's techniques could have fixed Dorothy's diseased mitral valve in 1951, no surgeon could fix an ASD. A doctor wrote in her record: "She is no candidate for surgery."

Not yet, thought Lillehei.

Dorothy was in and out of the hospital over the next few months, and Walt stopped by whenever he could. They talked of many things—the news, the weather, life on the hospital ward, which Lillehei now appreciated as few physicians could. Clarence Dennis had left Minnesota, but other heart research was under way, and Walt filled Dorothy in. The progress gave her hope.

It was nearing midnight on July 20, 1952, when an orderly delivered the draped body of a recently expired patient to University Hospital's autopsy room, a windowless space in the basement. Before beginning, the pathologist telephoned Lillehei. Lillehei had taped a note to the wall asking to be notified whenever there was an autopsy; night or day, he would come. It was the best way he could think of to learn the interior of the human heart.

The pathologist opened the body and started removing and weighing the organs. He gave Lillehei the one he wanted.

Lillehei cupped Dorothy Eustice's broken heart in his hands.

What tragedy, he thought. A lovely young woman like her, beaten by a simple hole.

Lillehei took Dorothy's heart to a side table and cut it open. With a few stitches, he closed the defect.

Any seamstress could have sewn that up in five minutes. If only we could get inside the living heart.

And that was the moment Lillehei dedicated himself entirely to open heart research. The question was, What could he bring to the quest? By the summer of 1952, the great minds that had been at Broadmoor seemed stymied.

OWEN WANGENSTEEN

5

Animal Testing

WHEN IT CROSSED his desk that summer of 1952, the *British Journal of Surgery* seemed unlikely to tantalize Walt Lillehei. The most promising heart research occurred in America, not Britain, where a powerful anti-vivisection movement dating from Queen Victoria's reign hindered animal experimentation. Without live animals, Lillehei knew, you got nowhere.

Like Victoria's fading empire, this latest issue of the *British Journal* was uninspiring; reading its pages, you had no sense that a great new epoch might be near. Lillehei scanned through workmanlike articles on cancers of the rectum, penis, and muscle. He read about a new procedure for amputating the human hindquarter, which is the leg from the hip down. Owen Wangensteen would be interested, but not a young surgeon intent on the heart.

Lillehei was nearly through the journal when an article stopped him. Only four pages long, it had an intriguing title: "Experimental Cardiovascular Surgery."

Written by scientists in Kent, England, the paper summarized experiments in closing off the blood supply to the hearts of laboratory dogs. Many studies, including Lillehei's, had already demonstrated how sensitive the brain especially was to oxygen deprivation—how after about four bloodless minutes, the damage was increasingly severe and irreversible, and shortly thereafter, the brain began to die. As experiments went, these were straightforward. A scientist anesthetized a dog, opened its chest, tied off the main vessels that return blood from the body to the heart, and, with an eye on a stopwatch, untied the vessels after a predetermined period. Then the scientist sewed up the chest and monitored the dog's behavior after waking. If all was well, the next dog went thirty or sixty seconds more without blood. No dog had ever survived beyond ten minutes.

But until the English scientists, no one had discovered the role of a tiny vessel called the azygos vein, which delivers a trickle of blood into the heart. The Englishmen found that if they left this tiny vessel open while everything else was clamped off, its miniscule flow (through the heart to the lungs and then back out the aorta to the body) was enough to sustain life—without brain damage, without any damage anywhere—for two hours or more.

Two hours! thought Lillehei. Remarkable.

If what the Englishmen asserted was true, it meant that a fundamental assumption of just about everyone working to perfect a heart-lung machine was not really fundamental. Clarence Dennis, John Gibbon, and others assumed that the brain and other organs required a normal flow of blood to avoid damage. Yet normal flow, as Dennis's operation on Patty Anderson had tragically shown, could flood the open heart, making visibility difficult and surgery treacherous.

The implications of the British study for open heart surgery were staggering. They meant success in the quest might be far simpler than anyone had imagined.

These were invigorating times for heart researchers.

Heart disease was becoming medicine's cause célèbre—as prominent as AIDS would later be. The American Heart Association had emerged as a powerful lobbyist, and President Truman had named heart disease, which killed more than 625,000 young and old Americans annually, "our most challenging public health problem." In 1950, when proclaiming February national heart month, Truman issued a call to arms. "Measures to cope with this threat are of immediate concern to every one of us," he said.

Scientists benefited by getting more money. The federal government was exceedingly generous with its grants—and in Minnesota, Owen Wangensteen was hardly shy about having his hand out. Nor was he restrained with his hiring.

Early in Wangensteen's tenure as chief of surgery, a promising young doctor had applied for a residency. Wangensteen wanted him, but didn't have the money to bring him on. So Robert Gross went east to Harvard, where, seemingly overnight, he became the foremost early pioneer in closed heart repair. Wangensteen vowed never to turn down a young fellow like that again. By tirelessly campaigning for federal funds, private contributions, and support from the Minnesota legislature, which set the University of Minnesota's appropriation, Wangensteen had increased his own department's budget from a measly $20,000 a year in the early 1930s to more than a million dollars twenty years later.

University Hospital bustled with Wangensteen's protégés. Unlike many of the professors back east, who worried that their disciples would one day overshadow them, Wangensteen delighted in his young men—at least one of whom, he was certain, would bring a Nobel prize to Minnesota. With space for surgical research scarce, Wangensteen opened labs on the top floor of Millard Hall, the domain of his friend Maurice B. Visscher, the chief of physiology. It was cramped up there, poorly lit, frigid in January and stifling in August—an attic.

Humans coexisted with caged mice, rats, cats, guinea pigs, hamsters, birds, rabbits, and the occasional monkey or chimpanzee. Cardiac surgeons favored dogs, not only because their hearts were uncannily similar to the human heart, but because they were plentiful and cheap.

Lillehei's lab looked like nothing special. It had two operating tables, operating lights, instrument cabinets, a sink, oxygen tanks, and two tables, all in less space than a caboose. Next door, separated only by a grate, was Claude Hitchcock, another surgeon. Hitchcock's interests were eclectic, even by Wangensteen's standards. Suspecting that the bodies of cockroaches manufactured a carcinogen, Hitchcock trapped the insects at the bottom of a hospital elevator shaft (where they'd eluded the exterminator), ground them up, and fed them to rats. Hypothesizing that the by-products of internal combustion played a role in lung cancer, Hitchcock mounted a lawnmower engine on a table, fired it up, and smothered other unlucky rats in exhaust. Researchers choked in noise and fumes.

Lillehei had no fancy machines in his lab. Clarence Dennis had by this time gone with his heart-lung apparatus to New York, where he was chief of surgery at a hospital in Brooklyn, and Wangensteen had suggested Lillehei pick up where he left off. Dennis had paid the university a few thousand dollars to take his machine, and Wangensteen recommended Lillehei use the money to build something similiar. But Lillehei declined.

I just don't think it's going to work, he said. It's too complex.

Wangensteen was surprised; despite Dennis's failures, almost everyone believed in some kind of machine.

But Lillehei had seen Sheryl Judge pumped full of air. He'd read about Ferdinand Sauerbruch, the German scientist whose solution to the problem of lung collapse during chest surgery had been the negative-pressure chamber, an outrageously expensive, labyrinthine, room-size contraption that for a few years had been all the rage on both sides of the Atlantic Ocean. Then researchers found something vastly superior to the chamber: a tube. A simple endotracheal tube, which cost less than a dollar.

And now another stroke of simplicity, from the *British Journal of Surgery.*

Well, it's your research, said Wangensteen. Do what you think is best.

With Dennis's money, Lillehei set out to confirm the Englishmen's azygos findings and see how they might apply to open heart surgery. Every day, after completing his regularly scheduled operations, Lillehei climbed the stairs to his attic lab, often working until after midnight. Two years had passed since his cancer operation, and Lillehei remained symptom-free—and now he knew what he wanted to do with his life, whatever was left of it.

That summer of 1952, he'd held Dorothy Eustice's heart in his hands.

———

To some on the outside, Millard Hall was evil incarnate. Who knew what unspeakable horrors took place in that attic? Were dogs being skinned alive, as one rumor had it? Was man's best friend being blinded, burned, or operated on while awake? Who would subject an innocent creature to such torture?

Only a devil, said Minnesota's anti-vivisectionists.

"I wish that savages like you would rot alive or be thrust into boiling water," one wrote in 1951 to Maurice Visscher, whose department occupied Millard Hall.

Another anti-vivisectionist put her faith in reincarnation: "I believe that some day you will come back to this earth as dogs and will be used for scientific purposes."

The anti-vivisectionists staged demonstrations and lobbied at the local, state, and national levels for an end to animal research. They pressured police to crack down on the supply chain, for dogs especially. Vendors claimed that they obtained their animals legally, but the truth was the less scrupulous among them stole. Anyone doubting that was referred to the front page of the December 17, 1939, *Minneapolis Tribune*. Police the day before had broken up a dognapping ring and the thieves had confessed to stealing some five hundred household pets and selling them to the University of Minnesota for $2.50 apiece. The cops had gotten on to the racket after the bereaved owner of a missing animal went inside Millard Hall.

"I saw at least 100 dogs there," said the owner. "Some had been operated on in medical experiments. And one of those dogs was my own. . . . They said they would heal him and return him to me." But the dog died, and the owner called the police.

Although many in Wangensteen's world considered the anti-vivisectionists irrational, no one took them lightly. The anti-vivisectionists were well financed. They had friends in Congress—and, until he died, in 1951, a powerful ally in the aging William Randolph Hearst, whose newspapers supported those seeking to end experiments on animals. The anti-vivisectionists had won many battles in England and some in America; in Minnesota in 1952, they were mustering their forces for a renewed fight to overturn the state's animal-experimentation law, which the legislature had passed, amid bitter controversy, in 1949.

And the anti-vivisectionists were not queasy. One photograph of a mangled dog—someone's beloved Fido—was devastating to the noble aura that scientists sought to create. Through paid advertisements and sympathetic reports in Hearst papers and other publications, the anti-vivisectionists presented horrifying images to the public.

In Minnesota, no one fought back as fiercely as Maurice Visscher.

An esteemed researcher who took strong public stands against militarism and anti-Semitism at a time when doing so was courageous (especially for a scientist), Visscher believed passionately that with many diseases, only animal testing could lead to saving human lives. At the University of Minnesota in the 1950s, Visscher maintained, scientists obtained their animals legally, fed them well, gave them adequate ventilation and space, and always anesthetized them before cutting.

Visscher was a gentle man, a Unitarian who ran a food bank out of the basement of his home, but he was no sissy. He did not back down when he had a cause, and none of his many causes consumed him like opposing the anti-vivisectionists. Visscher wrote letters to the editor, testified at hearings, and was the principal proponent outside of the state legislature of Minnesota's 1949 animal-experimentation law. Visscher went to Washington, enlisting the support of Hubert Humphrey and Eugene McCarthy, among others. He coaxed nervous scientists out of their labs and into the light of public debate, to prove they were anything but sinister fiends. He himself testified at hearings. Visscher's fight, begun in the 1930s, would last virtually to the day he died.

While he was too refined to call anti-vivisectionists loony, Visscher believed they were.

"I know of no persons of sound mind who doubt that medical research involving surgery of animals has greatly benefitted mankind in thousands of ways," he wrote to a congressman during World War Two. As for anti-vivisectionists, "they and Adolph [*sic*] Hitler may care to live in a world in which science is killed and medicine is manacled, but I am sure that most right-minded people will differ with them."

———

Although police had received a report of an anti-vivisectionist emptying a rifle into a scientist's home in California, animal-rights militancy in the early 1950s was rare. The anti-vivisectionists distracted; in Minneapolis, at least, they did not free animals or harass scientists, except in letters and at hearings and demonstrations. And so, the surgeons in Millard Hall persevered.

Among them was F. John Lewis.

Lillehei's medical school classmate, Lewis in 1952 was one year his friend's senior on Wangensteen's staff by virtue of the time Lillehei had lost to lymphatic cancer. If anyone working for Wangensteen qualified as a Renaissance man, Lewis was the one. He painted beautifully—still lifes and illustrations of operations. He was conversant on such diverse writers as Joyce and Kafka; during wartime service as an army doctor in the European theater, he not only compiled an extensive list of great books but also managed to read most of them. Lewis worked with wood, played with computers, hunted, fished, bicycled, and climbed mountains. Along with hundreds of medical articles, he wrote essays and poems and, later in life, after he'd left surgery altogether and moved to Santa Barbara, two published books. In one of his essays written after he put down the knife, Lewis looked back on the post-war 1940s and early 1950s, when he and Lillehei were just getting started on their medical careers. In his "imagined" surgeon, Lewis portrayed some of himself and his friend:

"This imagined or fabulous surgeon wanted to drink more, party more, and stay up all night talking while everyone else listened; thus, he was commonplace, but deeper. He had a spark of ambition and a sense of urgency brought on by the lost years of the War. . . . When opportunities appeared, he had a way of being in the right place with people whose motives he could understand. His purposes fit the general mood of the times."

During their medical school days, Lewis had passed many a fine drunken hour with Lillehei at Mitch's, the Dixieland joint that jumped until dawn—and now the two surgeons were regulars at the Parker House, which had succeeded Mitch's after the old bottle club had been torn down to build a highway. Over martinis—between dinners of lobster tails and steak, and dancing with their beautiful young wives— Lewis and Lillehei discussed ways to get inside the living heart.

One possibility especially intrigued Lewis: hypothermia, the cooling of the body for medical purposes. For several months now, Lewis had been experimenting in the dog lab.

Lewis had not come up with this idea on his own. He'd been sitting with Lillehei at Broadmoor in 1950 when Wilfred G. Bigelow, as yet an obscure Canadian surgeon, had presented a paper on the results of his experiments with the cooling of animals.

Hypothermia was not new even to Wilfred Bigelow—cold had been tried as an anesthetic and as a controversial treatment of the mentally ill. Indeed, some doctors had tried cold in an attempt to cure cancer, and Wangensteen one day would use it against his old adversary, the peptic ulcer. But those were fringe practices; mainstream medicine considered cold the enemy of tissue. Bigelow understood this firsthand: in 1941, early in his career, he had amputated a man's fingers after frostbite had turned them gangrenous. It was not an uplifting experience (especially when the man, a superstitious sort, demanded to take his severed fingers home), but Bigelow's curiosity was piqued and he read everything he could find about the effects of cold on living cells.

Inhumane though it was, one source was research conducted in Nazi Germany.

Seeking to improve the survival rate of *Luftwaffe* flyers who'd gone down in the icy North Sea, German scientists had experimented on concentration camp inmates. One who survived, the Reverend Leo Miechalowksi, a Roman Catholic priest, later testified at the Nuremberg trials.

The priest told of being brought to an experimental station—the so-called aviation room—at Dachau. After stripping Father Miechalowski naked, the Germans taped one set of wires to his back and inserted another set into his rectum, then fastened an inflatable collar around his neck. Then they immersed the priest in a basin filled with ice and water.

"I began to tremble," the priest testified at Nuremberg. "I immediately turned to these two men and asked them to pull me out of the water because I would be unable to stand it much longer. However, they told me, laughingly, 'Well, this will only last a very short time.' "

Father Miechalowksi remained conscious in the water for an hour and a half, during which time his body temperature slowly dropped from 37.6 degrees Centigrade, about normal, to 30 degrees. Every fifteen minutes, the Germans took a blood sample from the priest's ear. They made him smoke a cigarette, sip a glass of schnapps, and swallow a cup of grog.

"Now my feet were becoming as rigid as iron, and the same thing applied to my hands, and later on my breathing became very short. I once again began to tremble and afterwards cold sweat appeared on my forehead. I felt as if I was just about to die . . . then I lost my consciousness." Sometime later, Father Miechalowksi awoke on a stretcher, his body warmed by lights.

"Do you still suffer from the effects of this experiment?" Father Miechalowksi was asked at Nuremberg.

"I still have a weak heart," the priest said.

Barbaric though they'd been, the Germans had learned a lesson with potentially therapeutic implications: the human body can be remarkably tolerant of cold.

———

Bigelow had no intention of experimenting on unwilling human subjects. Nor, at first, could he think of any practical new use for hypothermia. He'd only learned that as metabolism slows, the demand for oxygen decreases—and that while freezing can kill tissue, chilling does not, necessarily.

In 1946 and 1947, Bigleow studied with Alfred Blalock at Johns Hopkins University. Bigelow watched Blalock perform his blue-baby surgery many times and was both impressed by Blalock's ingenuity and daunted by the limitations of the closed heart approach. "I became aware that surgeons obviously would never be able to correct or cure [many] heart conditions unless they were able to stop the circulation of the blood through the heart, open it, and operate in a bloodless field under direct vision," said Bigelow.

One night in Baltimore, the young Canadian awoke with an idea: Why not cool the whole body, reduce the oxygen requirements, interrupt the circulation, and open the heart?

For the next several years, back in Toronto, Bigelow experimented with dogs—120 dogs, when all was said and done. He learned that shivering impeded the operation, so first he anesthetized his animals. Then he buried them in ice, and, monitoring their vital signs, opened their chests and shut off the blood to their fading hearts. Sometimes he opened their hearts—and found a relatively bloodless field, suitable for surgery. When he was done, Bigelow sewed his dogs up and immersed them in warm water to see if they would awaken; unless they'd been below about 20 degrees Centigrade, they did. Moreover, they showed no damage—even in their brain cells, which required significantly less oxygen when chilled. Carefully controlled, Bigelow now believed, cold was no enemy but a promising ally.

Some who heard Bigelow present his research findings at Broadmoor in April of 1950 considered him screwy.

But not Lillehei, whose career a month later was sidetracked by cancer—nor Lewis, who immediately went to work in his own lab in the attic of Millard Hall.

———

Meanwhile, having mastered closed heart repair of the mitral valve, the renegade Charlie Bailey was seeking to correct more severe anomalies, such as atrial septal defect, or ASD.

The ASD was a logical next step for any ambitious surgeon who had mastered closed heart technique. A hole on the inner wall (septum) between the heart's two upper chambers (atria), an ASD could be more easily accessed than many other defects. An ASD was not dangerously close to the heart's nerves, and it was usually a single hole, easy to sew, as Walt Lillehei had demonstrated with Dorothy Eustice's lifeless heart. Some of the more complicated defects, autopsy studies had shown, consisted of multiple holes, holes hidden behind folds of tissue, or holes that were too large for simple suturing. Some consisted of holes in conjunction with faulty valves and other problems that would take a long time to repair. Compared with those monstrosities, an ASD was almost inviting.

By the summer of 1952, reports were surfacing of surgeons who had successfully closed ASDs. Robert Gross claimed to have conquered the defect with his cumbersome atrial well, and Bailey himself had an invention: the so-called doughnut method, in which the surgeon pressed the outer wall of the heart down over the hole in the septum, then sewed the outer wall to the septum with a circle of stitches, thereby creating a "doughnut." As with Bailey's mitral valve operation, a surgeon sewing a doughnut worked inside a closed heart.

But cold seemed to promise a better way to repair an ASD than doughnuts or Gross's gizmo. "Several of the operations could be improved upon if they could be performed by an open technique under direct vision," Bailey declared. "Since the heart-lung [machine] is not sufficiently perfected for clinical use, it would seem logical to use hypothermia for this purpose."

Starting, as always, with dogs, Bailey experimented with different methods of cooling. Spooked, perhaps, by Nazi experiments, Bailey did not immerse his animals in ice water. But he fiddled with the cold chamber, a six-foot-long freezer chest, discarding it after discovering that it cooled too well, causing frostbite. He buried animals in cracked ice, which was effective but messy. Bailey settled on rubberized cooling blankets, designed to treat schizophrenics—who had not appreciated the gesture, nor been cured.

After determining that he could safely shut off the blood to the hearts of refrigerated dogs for up to twelve minutes, Bailey decided to try a person. He chose a thirty-year-old woman with a large ASD. He knew the size of the defect exactly, for he had previously tried to close it with his doughnut, but it had been too big.

On August 29, 1952, Bailey brought the woman back to his operating room at Hahnemann Medical College and Hospital, which forgave his early failures when his mitral valve operation became such a sensation. The woman was anesthetized and her temperature was reduced to 27 degrees Centigrade, about the point at which unassisted breathing ceases. Bailey then clamped off the vessels into the heart and opened the heart wall; the atria were relatively dry, and he could see what he was doing, which was a blessed thing.

Bailey sewed the hole and closed the heart wall in six minutes. Then he unclamped the vessels—and the heart went immediately into fibrillation, the uncoordinated pumping that is fatal unless defibrillated. Bailey could not reverse it, and the woman died on his table.

Investigation revealed bubbles in the woman's coronary arteries, which supply blood to the heart muscle. Somehow, air had gotten into the woman's bloodstream. Air—deadly to all tissue, as Clarence Dennis, operating on Sheryl Judge the year before, had tragically demonstrated.

———

Word of Bailey's attempt had not yet reached Minneapolis on September 2, 1952, when a five-year-old girl with a suspected ASD was put to sleep at University Hospital.

The daughter of traveling carnival workers, Jacqueline Johnson was an underweight child who had been sick most of her short life. Her heart was enlarged and her doctor, F. John Lewis, believed that without surgery she would soon die. Satisfied with his laboratory experiments, Lewis was ready to try hypothermia in a person.

Assisted by Lillehei, Varco, and other surgeons eager for a place in history, Lewis wrapped the girl in rubberized blankets and turned on the cold. Jacqueline's temperature at the start was normal. It fell slowly, taking twenty-five minutes to drop the first degree, and another ten minutes to drop another. Steadily lower it went, along with Jacqueline's pulse. After two hours and fourteen minutes, the girl's core temperature had reached 28 degrees Centigrade. Beating almost 120 times a minute when the refrigeration went on, her heart was now slower by half.

The blankets were removed and Lewis quickly opened Jacqueline's chest. With small tourniquets, he and Varco closed off the superior and inferior venae cavae and the tiny azygos vein—the vessels that return blood from the body to the heart. Then he clamped the pulmonary artery, which sends blood to the lungs, and the aorta, which returns the freshly oxygenated blood back to the body. No blood could get into or out of the girl's heart.

No blood moved anywhere inside Jacqueline.

Lewis cut through the wall of the right atrium and found barely a glistening of blood. He located the defect immediately and, thank heavens, it was exactly as diagnosed: an ASD. Lewis began to sew it closed.

Lillehei watched the clock. They were four minutes in. Four minutes, the point at which oxygen-deprived brain cells at normal temperature start to die.

Lewis tied his last stitch. Then he filled the atrium with saline solution to test his work—and saw a leak. He placed another stitch. This time, the closure was tight.

Five minutes now.

Lewis closed the wall of the heart and released the vessels he'd shut.

Five and a half minutes.

Jacqueline's heart began to resume its normal load, but reluctantly. Lewis massaged the heart until it had regained proper rhythm. Then he closed Jacqueline's chest and immersed her twenty-nine-and-a-half-pound body in a tub of warm water.

Lewis, Renaissance man, had found the tub—a Farm Master watering trough—in a Sears Roebuck catalog.

———

Eleven days later, history's first succesful open heart patient went home, cured. When he learned of it, Wilfred Bigelow was pleased with Lewis's accomplishment—but crestfallen that he himself had not been first. After so many years of research, Bigelow had been ready to attempt an open heart operation on a person for almost a year, but no cardiologist would refer him a child, and Bigelow did not think it wise to try a larger patient first. Bigelow worked at an adult hospital in Toronto; the city's children's hospital, across the street, was home of a competing cardiac surgeon who was experimenting with monkey lungs as a heart-lung "machine." The surgeon's cardiologist partner would not send a case across the street.

In Minneapolis, Lewis's operation drew accolades—and generated headlines.

"It opens the way for more such operations on the type of defect repaired," said the *Minneapolis Tribune* in late September, after Lewis had described the case at a scientific meeting, the usual forum to break research news. "But even more significant and exciting, it seems to give surgeons a method— long sought—of putting the knife into the live human heart, in plain sight and unclogged."

Tribune editorialists also enthused, for a somewhat different reason. Unlike Hearst papers the *Tribune* opposed the antivivisectionists, who hoped to overturn Minnesota's animal-experimentation law at the next legislative session; now the paper, sympathetic to Maurice Visscher, had a new weapon in the battle for public opinion. Fourteen of forty dogs had died during Lewis's hypothermia research, an editorial writer noted, in order that a child could live. "One child at the price of fourteen dogs is a remarkable bargain any way you figure it," stated the writer. "Minnesota should be proud that its laws enable its scientists to save human life this cheaply."

———

Having assisted Lewis, Lillehei knew firsthand what a watershed this was; if only Dorothy Eustice had hung on a bit longer, they might have saved her.

Nevertheless, Lillehei did not believe that hypothermia was the end of the quest. He doubted that hypothermia could ever be extended long enough to afford the time needed to fix more complicated heart problems.

He doubted it would ever allow a surgeon to repair a ventricular septal defect, for example, a defect involving the heart's lower two chambers—never mind such complexities as atrioventricular canal, the mysterious defect that Clarence Dennis had found inside Patty Anderson, nor the terrible tetralogy of Fallot, which Blalock's blue-baby operation remedied but did not cure. Many doctors in 1952 still believed that no matter how ingenious the technique or technology, no surgeon could ever fix either of those.

The *British Journal of Surgery*'s report on the azygos vein tantalized Lillehei—but were the Englishmen's findings really just another tease? There was only one way to find out.

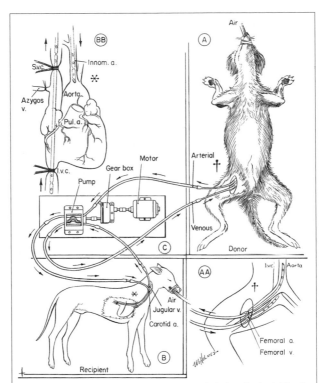

Fig. 1.—Controlled cross-circulation. *A*, Donor animal showing site of arterial and venous cannulations. *AA*, Close-up of the femoral arterial and venous catheters positioned with their tips in the aorta and vena cava, respectively. *B*, Recipient animal (patient) demonstrating the position of the arterial cannula in the carotid artery directed toward the aorta, and the venous cannula inserted through jugular vein and positioned so that venous blood is simultaneously removed from both the superior and the inferior venae cavae (refer to closeup, *BB*). *C*, The pump. The entire system has only two controls: one is the speed changer which couples the motor to the pump and which is set prior to starting the pump, and the other is a switch to turn the motor on and off.

CROSS-CIRCULATION IN THE LAB

6

A 10 Percent Solution

B Y THE TIME he assisted John Lewis at history's first successful open heart operation, Walt Lillehei had been promoted to associate professor at the University of Minnesota. He had not only free rein with his research but a full-time resident to run his lab, Morley Cohen.

A thirty-year-old Canadian, Cohen, like so many other aspiring surgeons, had been drawn to Minneapolis by the unusual opportunities Owen Wangensteen offered. And now, what better opportunity than to work for Wangensteen's prize pupil?

Under Lillehei's direction, Cohen had been experimenting with ways to oxygenate blood. Inspired by the tropical fish he'd had as a kid, he was monkeying around with aquarium bubblers purchased at a neighborhood pet shop. He could oxygenate dog blood with these air stones—but getting the bubbles out after they'd done their job was difficult, perhaps impossible. After weeks of trying, Cohen was frustrated.

Lillehei showed Cohen the article from the *British Journal of Surgery.*

Take a look at this, said Lillehei.

Cohen agreed that the lessons of the azygos vein—the azygos factor, they called it—seemed to have great potential. The job now was to measure the flow of that tiny vessel. The Englishmen had only demonstrated its importance, not produced hard numbers that might be applied to surgery.

Cohen got to work using the same technique as the British scientists, with one addition: after tying off the superior and inferior venae cavae, the primary vessels that return the blood to the heart, he collected timed samples of the much smaller volume of blood that continued to trickle through the azygos (his collection device was a condom, which lay nicely inside the chest cavity and didn't easily leak). Cohen used nineteen dogs, relating body weight to azygos flow. He found that an animal survived, unharmed, for thirty minutes or more with less than one tenth its normal blood flow.

Less than a tenth! It contradicted reason.

But there it was, demonstrated in research that was unimpugnable.

A brainstorm struck Cohen: Why not use a lobe of one of the dog's own lungs to oxygenate its blood? One lobe could produce the minimal flow of blood that the azygos factor called for—and living tissue would naturally oxygenate. Using nothing but a pump, some tubing, and the lobe, Cohen proposed to bypass the dog heart—to turn the animal itself into its own heart-lung "machine."

This was an ingenious yet simple idea, the kind Lillehei cherished. Walt told Morley to begin a new round of experiments on dogs. It was by now 1953. In a growing number of centers across America, the quest was intensifying.

If not for his machine, history might not have remembered John Gibbon. Gibbon was not an unusually fast or innovative surgeon, nor was he a universally beloved teacher. Some thought him aloof, perhaps not surprising given that he was a graduate of Princeton who'd trained in surgery at Harvard.

What meager introspection he seemed to allow revealed a man of little cheer, as evidenced by a poem he wrote:

The boxes line up row on row
In and out of boxes all our lives
The womb's a box
The coffin, too
Big concrete slabs pushing to the skies
Filled with boxes
Inside the box, softness lies here and there
To cushion our bodies in chairs and lounges
To hold our bodies in gentle sleep
As we coast down hill from cradle to the grave.

But Gibbon had his machine, long before Clarence Dennis or the many others who were building contraptions had theirs. Gibbon had started work in the 1930s, while he was at Harvard, and continued during his years as a professor of surgery at Philadelphia's prestigious Jefferson Medical College.

Although Gibbon himself shied from publicity, his heart-lung machine experiments became a public sensation—for in the popular mind, at least, few spectacles were as beguiling as man attempting to play God with electricity and steel. Moreover, Gibbon's work had attracted the attention of Thomas J. Watson, the chairman of IBM, an emerging power in the fledgling computer industry. The businessman provided the doctor money and technical support to help develop the heart-lung machine—which some said looked just like a computer.

"This robot, a gleaming, stainless steel cabinet as big as a piano, will soon be tested on humans," *Life* magazine proclaimed in May of 1950. "It already has successfully substituted for the living heart and lungs of nine dogs for as long as 46 minutes."

In fact, Gibbon did not try his machine on a person until almost two years later, when he operated on a dying fifteen-month-old girl who had been diagnosed with an ASD. Gibbon opened the baby's chest, connected her to his machine, then cut into the right atrium of her heart. The bleeding was substantial, but Gibbon could see tolerably well with the constant use of the suction. Nonetheless, he encountered a nightmare: exploring as best he could with eyes and fingers, he found no hole, nor any other abnormality inside the heart; the diagnosis of ASD had been wrong, and the baby perished. An autopsy revealed the true cause of her sickness: patent ductus arteriosus, the birth defect involving the heart's outside anatomy that Robert Gross had first fixed in 1938—and which surgeons had routinely fixed ever since.

"This of course illustrates the importance of complete exploration of every heart which is operated on," said Gibbon. "We might have saved this child's life if we had closed the ductus."

Although he was satisfied with how his machine had worked, Gibbon did not try again until May 6, 1953, when he operated on Cecelia Bavolek, an eighteen-year-old whose failing heart had sent her to the hospital three times in the last six months. Bavolek had also been diagnosed with an ASD, which is what Gibbon found when he opened her heart and deployed a steady suction that allowed him to see, if only "adequately," as he later wrote. During the twenty-six minutes that Gibbon's machine sustained the woman's life, Gibbon repaired the defect.

Thirteen days later, Bavolek was discharged from the hospital, a normal young woman.

Gibbon's success did not just overshadow John Lewis's landmark achievement with hypothermia; it seemed that in those twenty-six minutes, the Philadelphia surgeon had captured the Grail. The only task left was duplicating his results, so that mankind, not just one lucky teenager, would benefit.

But Gibbon's next two attempts, the following July, were disasters.

88

One of the patients, a five-year-old girl, was lost before Gibbon really even started: soon after he laid bare her heart, it turned blue, ballooned, and stopped beating. Gibbon forged ahead, massaging the swollen heart until its color returned, after which he connected girl to machine. Gibbon then successfully sewed up her ASD and closed the heart wall, but he could not wean her off the machine: every time he tried, her heart arrested. After some four hours, Gibbon finally gave up and disconnected the girl for good, and she died.

Gibbon's next try was on another five-year-old girl. Hoping again to find an ASD, as diagnosed, Gibbon opened the child's heart and did find that defect—and also two others, patent ductus arteriosus, and a hole between the lower chambers of the heart known as a VSD. Whether he could have fixed all three defects was now an academic issue, for the bleeding this time was as torrential as Clarence Dennis had encountered inside Patty Anderson, and the suction could not keep up.

"As we could not get a clear field to work in," Gibbon wrote, "and the flow of bright red blood was so excessive, we closed the atrium and removed the cannulae." The girl then died.

———

Gibbon was among the surgeons who traveled to Minneapolis two months later for one of Owen Wangensteen's surgical symposiums.

Readers had learned of Gibbon's one success in the May 18 edition of *Time* magazine, which declared that the Philadelphia surgeon had "made the dream a reality." But even in Wangensteen's audience, few knew about Gibbon's July disasters. Gibbon informed his fellow surgeons at the September symposium, and they assumed he would continue on; he was, after all, the father of the movement. Yet Gibbon was so discouraged that he declared a one-year moratorium on all further human use of his machine.

In fact, Gibbon, fifty years old, would never use it again; for him, the quest had ended.

Still, Gibbon did not act humbled at the symposium. He remembered all too well how one of Wangensteen's men, Clarence Dennis, had beaten him into the operating room with a heart-lung machine—a machine whose design could be traced to his own! And now, another young Turk from Minnesota claimed to be nearing a cardiac breakthrough.

Gibbon liked Walt Lillehei. The older surgeon had watched him deliver his first national paper at Broadmoor three years before; to an Ivy league–educated Philadelphian, the young Midwesterner with the alligator shoes and gold watch was too flashy; Lillehei seemed almost to swagger, as if, at the tender age of thirty-four, he'd already scaled tall peaks. Outside of surgery, about all Professor Gibbon had in common with Dr. Lillehei was an appreciation for fine liquor.

So when at Wangensteen's symposium Lillehei talked about the azygos factor, Gibbon dismissed it. He didn't care what Lillehei's experiments purportedly showed. This 10 percent notion—what was it called again, please?—was ridiculous.

It's just not possible, Gibbon said. Animals need much more blood.

Lillehei pulled Morley Cohen aside and said: Now I know we're right! Gibbon doesn't believe it!

———

Not long after Wangensteen's symposium, Cohen was performing another of his "self-lung" operations on dogs, in which a lobe of a dog's lung functioned as the dog's own heart-lung "machine."

By any measure, the self-lung series had been a success: of nearly fifty dogs whose hearts had now been bypassed with this method, more than 90 percent had survived. Cohen and a second resident Lillehei had brought on board—thirty-three-year-old Herbert E. Warden—had mastered the mechanics of the technique. And physiologically, the studies showed, the self-lung seemed kinder to blood than running it through an actual machine, with its many artificial components.

But no one was uncorking champagne; if anything, Lillehei and his residents were increasingly pessimistic. Unless the tubes were handled with extraordinary care, disaster loomed: one barely noticeable kink, and blood backed up into the lung, which quickly swelled and was ruined. A master surgeon might be able to avoid kinking every time, but an ordinary surgeon under all the other pressures of open heart surgery would be dangerously susceptible to error. Lillehei, Cohen, and Warden were beginning to wonder if the self-lung was too delicate a business ever to be used with patients.

On an autumn day in 1953, it happened again: a tube kinked. Cohen and Warden straightened it and continued with the operation. The dog was fine, but Cohen's mind was elsewhere. He'd recently learned his wife was pregnant with their first child. Warden had already offered congratulations.

And that's when Warden's brainstorm struck: there in the attic of Millard Hall, his hands inside a mongrel dog's chest.

Wouldn't it be nice, said Warden, if you could plug a patient who needed an open heart operation into something like a placenta?

It would, wouldn't it, said Cohen.

Where could they get a placenta—the organ that sends life-sustaining blood from mother to fetus?

From a second animal, of course.

But why go to the bother of a placenta? Why not just plug into an entire second animal—an animal "donor"?

Cohen and Warden were excited, as was Lillehei, who knew about doctors who had treated leukemia by connecting a patient to a donor and letting bad blood mingle with good. A search of the literature disclosed a surprising number of attempts to treat other disorders, including cancer and high blood pressure, with cross-circulation. Someone had even tried it to bypass animal hearts. All of these scientists owed a debt to the nineteenth-century French physiologist Charles

Edouard Brown-Sequard, who had briefly restored eye and facial movement to the heads of decapitated dogs by injecting freshly oxygenated blood into the severed arteries of the cut-off heads.

But cross-circulation, in whatever form, had never succeeded in human use. In the fall of 1953, no one seriously believed it had a future in cardiac surgery, or anywhere.

No one but Cohen, Warden, and Lillehei, who didn't wonder if the glass was half empty or half full, but whether it was the right vessel in the first place.

———

Lillehei and his residents sketched out their approach; as always, simplicity guided them.

Blood flow would have to be gentle and precise, so they needed an easygoing pump that could be calibrated. They chose a $500 Sigmamotor Model T-6S, used in the dairy industry. The Model T-6S could simultaneously move separate streams of liquid in different directions, in exactly equal amounts—almost like the human heart. It was a sturdy machine with a soft touch: it did not kick up foam, the tiny bubbles that are deadly to living tissue.

Traditional gum rubber would not do for the tubing, for it was opaque—air bubbles could hide, and a surgeon could not see that a flow was being maintained. Lillehei went to Mayon Plastics, a Minneapolis company that was run by an old high school classmate, and found what he wanted: clear beer hose, used to connect barroom taps to kegs.

And that was about it, a milk pump and beer hose.

One day in October, Warden and Cohen anesthetized two dogs. They shaved a leg of the donor dog and made small incisions to expose an artery and a vein; short, rigid tubes—cannulas—were placed in each. Then they opened the chest of the "patient" dog and placed cannulas into that dog's appropriate artery and vein. The cannulas were tied into the beer hose, which, via the Model T-6S, connected the patient to the donor. The circuit was complete.

Calibrated to deliver the minimal flow the azygos factor called for, the pump was turned on and the patient's heart was clamped off. For half an hour, donor dog supported patient dog. Warden and Cohen then stopped the pump, removed the tubes, sewed the dogs up, and let them both awaken.

A mistake in arranging the tubing proved to have caused brain damage in the patient dog, but the next sixteen dogs recovered without harm. Thirty minutes of potential open heart operating time, and no damage—none to the donors, either! This, *this* might be the Grail.

Almost as remarkable was how quickly the dogs awakened. Lillehei well remembered Clarence Dennis's animals, across the hall there in the attic of Millard Hall: how lethargic they'd been waking up after only a few minutes on Dennis's heart-lung machine, how it seemed the very act of blood flowing through a complicated apparatus had been damaging, through some physiological reaction as yet unknown.

Cross-circulation offered other advantages. Cleaning and sterilizing was troublesome—a machine didn't fit into an autoclave, and chemical solutions didn't always kill 100 percent of the germs. But beer hose was easily sanitized and, since it cost next to nothing, a clean new length could be used every time. And cleaning and sterilizing the pump wasn't even a consideration, for no part of the T-6S came in contact with blood. The beer hose was placed behind moving metal fingers, which pushed the blood along.

Directed by Lillehei, Warden and Cohen experimented with different flow rates between donor and patient dogs. They analyzed cross-circulated blood for changes in oxygen, carbon dioxide, pH, hemoglobin, and other vital factors, and they measured blood pressure and pulse. After experimenting, they killed the animals and microscopically examined their livers, kidneys, hearts, lungs, and brains. They found no evidence that cross-circulation caused any damage.

As the autumn of 1953 progressed, Lillehei approved a second set of experiments. His assistants opened a dog's heart and, with an ordinary laboratory cork bore, made a large hole in the wall between the lower chambers, the ventricles—a ventricular septal defect, or VSD. Then they sewed up the defect. Their purpose was less to become technically accomplished in the repair (it was only sewing) than to confirm that the low flow rate of the azygos factor provided a semidry field—that it provided visibility for the surgeon.

It did.

Lillehei and his assistants also sought to determine how the canine heart tolerated such trauma.

The heart tolerated it very well.

Lillehei was by now convinced that cross-circulation would allow him to repair a human VSD—and with experience, atrioventricular canal, tetralogy of Fallot, and other extreme defects that remained beyond the surgeon's grasp. Just to be sure he hadn't missed some subtle yet dangerous neurological complication, Lillehei decided first to experiment on trained animals, whose behaviors were well known and could be more reliably tested post-operatively than pound dogs or strays. Lillehei asked Paul F. Dwan, a cardiologist friend, to donate some of his purebred golden retrievers.

Son of a founder of 3M, the Minnesota Mining and Manufacturing Company, Dwan was a millionaire—but money had not brought him health. As a child, Dwan suffered from rheumatic fever, which permanently damaged the valves of his heart and left him vulnerable to pneumonia and heart failure for the rest of his life. During one of the many times he was hospitalized, Dwan, then a young pediatrician, decided to pursue cardiology—and thus became, in the 1930s, the first pediatric cardiologist in the state of Minnesota and one of the first in the world. A professor at the University of

Minnesota when Lillehei met him, millionaire Dwan worked for free.

Dwan used his golden retrievers for hunting, but when Lillehei asked to experiment on some of them, the pediatric cardiologist not only agreed—he offered to have his trainer put them through their paces after they'd been subjected to cross-circulation. Lillehei accepted Dwan's offer and found that, after cross-circulation, the purebreds performed impeccably, as usual.

Lillehei began to look for a human candidate.

———

When Frances Glidden learned she was pregnant, in the summer of 1952, she and her husband, Lyman, sometimes thought of their daughter LaDonnah.

Donna, as they called her, was a skinny, freckled girl whose round face and red hair evoked her father's Irish ancestry. Donna was carefree and pretty. She liked cats and dogs and the family goat especially. If you needed Donna, the first place to look was out in back, where the Gliddens gardened and kept their many animals.

They lived in the north woods of Minnesota in a house not much bigger than a cottage. Lyman was a miner who worked the Mesabi Range, some ten miles north, in Hibbing. Over the decades, steam shovels had dug the world's largest hole in this fabulous iron lode—a pit three miles long, one mile wide, and as deep as 435 feet. Mesabi ore, by one estimate, was the raw material for a quarter of all the steel manufactured in post-war America.

At the beginning of 1950, twelve-year-old Donna was the picture of health.

A doctor two years earlier had detected a murmur and a marginally enlarged heart, but this seemed benign: Donna played sports, did tolerably well in school, and made her First Communion. Her appetite and color were normal, and her only sicknesses were the common cold and the flu.

Then, in the spring of 1950, Donna's energy suddenly flagged. It was if some unseen enemy had sucked the life force out of her. Her knees ached and she could not walk three blocks without becoming exhausted. She had trouble breathing; once, while in the yard, she fainted dead away. The doctors in Hibbing sent her to University Hospital in Minneapolis, where a cardiologist performed a heart catheterization—and diagnosed a ventricular septal defect, a VSD. The cardiologist could do nothing for the girl except prescribe digitalis and place her on a low-salt diet.

Donna was hospitalized in Hibbing for a week that September, but she was much better when she was discharged on Thursday, September 14. She went to bed that Friday night excited about a birthday party she was to attend that weekend. Donna shared a bed with her younger sister Shirley, one of the Gliddens' ten children, and they often drifted off to sleep with their arms around each other. Sleep came easily that night: the weather was still summery, and a warm breeze reached the girls through an open window.

When she awoke the next morning, Shirley played in their bed with one of the kittens. She thought her sister was still asleep. Then their mother came into the room. Something about the way Donna was lying alarmed Frances Glidden.

Frances called to her husband.

Lyman touched Donna and knew immediately that she was dead.

Now, two summers later, Frances was again expecting a baby. Fall passed and it was an uncomplicated pregnancy, as LaDonnah's had been.

Gregory Glidden was born on February 24, 1953. His color, cry, and appetite were all good. Just like his sister LaDonnah, he went home a healthy infant.

GREGORY GLIDDEN (*RIGHT*) WITH SIBLINGS

7

The Human Candidate

A T FIRST THERE was a stubborn fever. When it hadn't broken after four days, Frances and Lyman Glidden brought their baby to Hibbing General Hospital. It was April 7, 1953, and Gregory was just six weeks old.

There's no reason to worry, the Gliddens tried to reassure themselves. Babies catch colds so easily. Just because Donna had a bad heart doesn't mean Greg does, too.

A doctor diagnosed bronchitis; Gregory's heart appeared to beat normally, and the doctor heard no murmur. Antibiotics were prescribed, and after six days, Gregory was well enough to go home. The Gliddens were relieved.

Not two weeks later, Gregory spiked another fever and required rehospitalization. Again, antibiotics worked and the doctor pronounced Gregory's heart normal. After eleven days, the boy went home again.

By now, the Gliddens were worried. They knew that congenital heart defects could go undetected early in life—just look at their Donna, who made it ten years without a clue that she was doomed. In quiet moments, Frances put her ear to her son's tiny chest. After ten children, she certainly knew the sound of a normal heart.

What she heard inside Greg beginning in the middle of May chilled her.

It was the sound of Donna's murmur.

Frances told the doctor about it at Gregory's third hospital admission, on May 20, for yet another episode of bronchitis. The doctor this time heard a murmur, too—"loud, blowing, systolic," he wrote in Greg's burgeoning record. He ordered a chest X ray, one of the few tools a small hospital in northern Minnesota had in 1953 to diagnose heart problems. The X ray showed an enlarged heart, and the radiologist concluded: "The findings either represent an inter-atrial or inter-ventricular defect."

An ASD or a VSD.

A hole, like the one that killed Donna.

Francis and Lyman were distraught. Common sense suggested that, discovered so early, their son's heart disease would take him quicker than Donna's had. His lungs were obviously already damaged, and Greg, not yet three months old, seemed worn out. "Baby weak," a nurse noted. "Gets very tired when he eats," another observed. Gregory was propped up with a pillow, which helped him breathe, and placed in an oxygen tent. This time, he did not go home for fifteen days.

Only a week later, he was back in the hospital for a stay that extended to a month. B. F. Flynn, the family physician, said there was little they could do for Gregory in Hibbing, but Minneapolis might be another story. Flynn knew that John Lewis was fixing ASDs under hypothermia, and that surgeons in Owen Wangensteen's group were developing other promising techniques. Flynn made the necessary arrangements and, in September of 1953, the Gliddens drove south.

The Gliddens had been to University Hospital only once, with Donna, three years before. The hospital seemed different this time. Maybe it was wishful thinking, but the doctors sounded more knowledgeable, as if some of the mystery surrounding congenital heart disease had been peeled away. It seemed as if there were more heart doctors, and, in fact, there were. Pediatric heart surgery and pediatric cardiology were emerging as recognized specialties.

What was also different was the Variety Club Heart Hospital, University Hospital's new wing—a $1.6-million, seventy-eight-bed hospital on the banks of the Mississippi River, the first of its kind in America.

The hospital was inspired by a Minnesota businessman who himself had incurable heart disease, and who was an officer of the Variety Club, a charity whose founding in Pittsburgh in 1928 had made coast-to-coast headlines. On Christmas Eve of that year, an infant girl was found abandoned at Pittsburgh's Sheridan Square Theater. A note had been left with the infant. "Please take care of my baby," wrote its author, who identified herself only as "A Heart-Broken Mother." "Her name is Catherine. I can no longer take care of her, I have eight others. My husband is out of work." When police were unable to track down the woman, the baby was adopted by a group of theater owners and others in the entertainment industry who had recently established a small charity, the Variety Club. Spurred by all the publicity surrounding little Catherine, whose story prompted an outpouring of donations, the local charity grew into an international organization.

The Minnesota businessman was dying of his cardiac disease, but he succeeded in starting a drive, in 1945, for money to build a heart research and treatment center in Minneapolis. The cause was instantly popular—along with polio, heart disease was a national scourge. The businessman's local branch of the Variety Club raised a third of the cost of the new facility, with the remainder coming from federal, state, and private sources. "This could be called a hospital of plaques. They are placed in every room, above each bed," noted the program at the hospital's dedication, a gala evening in March of 1951 that featured an appearance by Oscar-winning actress Loretta Young.

What a marvel, the Variety Heart Club Hospital—"a doorway to heart health," the official brochure declared. The four-story hospital had a ward for children, a separate ward for adults, a floor for research, and an outpatient clinic. Unlike many other hospitals of the time, where the psychological needs of sick children went unmet, this hospital had ample space for play and learning, and a modern theater with 35mm and 16mm projectors, room for wheelchairs, and a stage for guest performers. The designers even understood the importance of color: in place of institutional white, rich browns, reds, and greens predominated in the Variety Club Heart Hospital. In winter, a fireplace warmed the lobby.

Frances and Lyman Glidden did not know it yet, of course, but this building on the banks of the Mississippi River would soon be their son's home.

———

Dr. Ray C. Anderson, one of Lillehei's cardiologist associates, conducted the initial workup of Gregory Glidden on the baby's first visit to the Variety Heart Hospital clinic in September of 1953. Anderson took the family history, examined the baby, listened to the murmur in his tiny chest, and studied new X rays that confirmed an enlarged heart and suggested an abnormal flow of blood between the chambers. An electrocardiogram of Greg's heart was also abnormal.

"It was my impression that this child has congenital heart disease," Anderson wrote to B. F. Flynn, Greg's doctor in Hibbing, "and that the most likely defect is an inter-atrial septal defect."

With his hypothermia operation, Lewis could fix an ASD. But Anderson could not be sure that Gregory did not have the more intimidating VSD, which no one could fix.

More extensive testing was needed. Anderson asked the Gliddens to bring their son back the next month for a heart catheterization, the best test for cardiac disease that medicine in 1953 could offer.

———

Walt Lillehei always appreciated the story of Werner Forssmann, the doctor who attempted the first human heart catheterization, in which pressures and oxygen content inside the heart are measured as a means of diagnosing cardiac disease. It was another demonstration of how sometimes only bold strokes advance the state of the art—and of how the best innovations often are greeted with envy, skepticism, and worse. As Lillehei's own surgical career progressed, he would come to know such responses well.

Forssmann was a doctor in 1920s Berlin, "a rather queer, peculiar person, lone and desolate, hardly ever mingling with his coworkers socially," said a colleague. Seeking a new way to deliver medication directly to the heart, Forssmann began to experiment with cadavers. He learned that a catheter—a long, narrow, flexible tube—inserted into a vein near the elbow could be easily threaded up the arm, then down the chest into the heart. It sounded Frankensteinian, but it worked.

Finding no volunteers on whom to experiment, Forssmann tried the procedure on himself. Without anesthesia, he opened a vein near his elbow, inserted a catheter, and began to inch it forward. Unable to take any more, Forssmann's assistant begged him to stop. For the assistant's sake, Forssmann did.

A week later Forssmann tried again, alone. Once more, without anesthesia, he opened a vein and worked the catheter toward his heart. He later wrote that he felt no pain: "I only perceived some sensation of warmth. There was also a minor stimulus to cough."

Having threaded the catheter as far as it could go, Forssmann needed an X ray to determine whether the tip had reached his heart. He climbed upstairs to the radiology department and took one. Then he found a bed, lay down, and nearly passed out. Fellow doctors thought at first he'd attempted suicide—he seemed the type. They found him, catheter sticking out of his arm, clothes and bed linen bloody, staring silently at the ceiling. Heart catheterization had worked, but years would pass before it became widely accepted. Most doctors initially considered it bizarre and dangerous, its lone proponent a mad German. The detractors never imagined that Forssmann one day would win the Nobel prize for his invention.

———

There were no CT scans or MRIs in 1953, no electron-beam-computed tomography—none of the tools that have since made diagnosis of heart disease computer precise. About all a doctor had was a stethoscope, X ray, electrocardiogram, and, now, cardiac catheterization, which measured blood pressure and oxygen content, reducing the margin of error but not eliminating it. As Dennis, Gibbon, and many others knew all too well, a conclusive diagnosis (outside of autopsy) came only when the heart of the patient lay open on the table.

Three weeks after their first visit to Variety Club Heart Hospital, Frances and Lyman Glidden returned with their eight-month-old son. They brought Greg into a room on the clinic level of the building, and then they were asked to leave.

As a nurse sang in an attempt to soothe him, Gregory was laid on a table and bound, like a mummy, with Ace bandages. Yet another of Lillehei's cardiologist friends, Paul Adams, inserted a thin catheter into a vein in Gregory's right arm and, guided by X ray, pushed it slowly toward his crippled heart. Blood samples were drawn and analyzed. The catheterization took about an hour, after which the cardiologist called Gregory's parents back into the room.

We think it's a VSD, said Paul Adams.

He made a sketch: it depicted a large hole in the wall between the ventricles, the two lower chambers of the heart. Three years ago, a doctor had made a similar sketch of Donna's dying heart.

It's not something Dr. Lewis can fix, Adams said. Nor can anyone else.

But Adams held out hope. He told the Gliddens about a colleague of Dr. Lewis's, a Dr. Lillehei who was hard at work in his laboratory on a way to give a surgeon enough time inside the open heart to fix a VSD. The experiments were going so well, Adams said, that Dr. Lillehei expected to be able to try the operation on a person in the very near future.

The Gliddens were cautiously encouraged.

———

Gregory wasn't home long when he was hospitalized yet again, in Hibbing, for bronchitis. Although he'd gained no weight since October, X rays showed that his heart had continued to enlarge, a sign of worsening failure (forced to pump harder, the heart grows extra muscle, a protective response that is eventually self-defeating).

It was the middle of December, the start of flu season and another long prairie winter.

Down in Minneapolis, Walt Lillehei was ready to go into the operating room with his new technique. All he lacked was the patient.

In an ideal world, Lillehei's candidate would have been a boy or a girl with a heart defect who was still in relatively good health; such a child would be more likely to survive the trauma of anesthesia and open heart surgery than one who was close to death. Yet how could Lillehei justify subjecting a comparatively healthy child to radical, untried surgery when, by doing nothing, the child might comfortably live a decade or more? If doctors had learned anything about heart disease, they'd learned how unpredictable it was.

Still, a sick child stood a greater chance of dying during surgery, and Lillehei feared the repercussions—both inside and, potentially, outside the University of Minnesota—if he lost a patient. Some naysayers were already unnerved by the goings-on up there in the attic of Millard Hall.

Another factor raised the stakes even further. During cross-circulation, the patient's life would be supported by a second, healthy person—the donor. The bloodstreams of patient and donor would be connected by tubing placed into arteries and veins, and the donor's heart and lungs would substitute for the patient's while the patient's heart was opened. Like the patient, the donor would be subjected to the risks of anesthesia, and any number of other problems could arise. Theoretically, both patient and donor could die.

But Walt Lillehei did not recoil from death. He'd survived World War Two and Wangensteen's knife.

After weighing everything, Lillehei cast his lot with Gregory Glidden. Gregory's parents had already told Dr. Adams that they would consent to surgery if it offered any chance of saving their desperately ill baby.

And time seemed to be running out.

Another of Gregory's doctors in Hibbing, George Erickson, had just sent an urgent letter to University Hospital. Gregory was rapidly deteriorating—he was coughing constantly now; he could not draw an easy breath; and with a temperature of 103, he was sweating profusely. Gregory had no appetite, and what little food the nurses got into him he vomited. To add insult to injury, the large toe of his left foot was infected, Lord knew from what.

"We were wondering if you have anything further to offer at this time," Erickson wrote to the university doctors. "Any help in this case would be greatly appreciated."

Shortly before Christmas, Adams wrote back to Erickson: "Our plan for this patient was to consider him for our first operation in January using the artificial heart that Dr. Lillehei has developed here."

But in light of Gregory's condition, Adams said, they would admit him as soon as possible—on Monday, December 28.

106

That December afternoon, Gregory was carried into Variety Club Heart Hospital in his father's arms. Lyman and Frances had come straight from Hibbing General Hospital, several hours north, where their baby had spent his first Christmas. Frances had not visited often during that stay. It was just too much for her: seeing Greg brought back the memory of Donna.

Lillehei introduced himself not long after Gregory was settled in. To the Gliddens, Lillehei didn't seem old enough to hold the power of life and death in his hands. He was only thirty-five, and he looked more like a Hollywood actor than a surgeon. The scar that ran from his left ear down his neck barely detracted from his looks.

But Dr. Lillehei didn't act like a star. The Gliddens had found many doctors to be rushed and apparently disinterested, as if anxious to knock off for the day—but not Lillehei. He really listened. He seemed unusually empathetic, as if he himself had once been at the mercy of medicine.

No, he told the Gliddens, there are no guarantees. But if a child of his were to need open heart surgery, he would not hesitate to use cross-circulation.

Lillehei told the Gliddens that his "artificial heart" was actually another person—in fact, one of them, if their blood type was compatible with Gregory's.

The surgeon drew a diagram of cross-circulation and talked of his experimental success with dogs, whose hearts were so close to humans'.

I think we can fix Gregory, Lillehei said.

The Gliddens needed no further persuasion. So they signed University Hospital Form 33, which consisted of a single sentence: "I, the undersigned, hereby grant permission for an operation or any procedure the University staff deems necessary upon my son, Gregory Glidden."

Lillehei's immediate objective was to stabilize his new patient in preparation for the surgery.

The amazing thing was, he was such a good baby. When he wasn't in pain—and there were days when he wasn't— Gregory had a wonderful laugh. He had tousled brown hair and beautiful eyes, and he was endearingly scrawny.

After an accidental overdose of digitalis that nearly killed him shortly after admission to University Hospital, Gregory improved. Developing a taste for Jell-O, pudding, and chocolate milk, he put on weight. His breathing eased, his fever broke, a hint of color came into his cheeks, and he began to sleep through the night. If was as if the curse upon him had been lifted. Heart disease was truly unpredictable.

Greg seemed not to long for his parents, who had returned to Hibbing. "He is cheerful and happy and kicks and plays with his toes in his crib," one nurse wrote into his record. "Laughing and waving hands about," wrote another. "Has been playing and cooing most of the evening."

Weeks passed and the nurses grew to adore Greg. He played in his crib with his stuffed animals and toys, but what he liked best was company. The nurses held him whenever they could, and they would park him in an infant seat on the nurses' station desk so he could be in the middle of the action. They read to him, pushed him up and down the halls in a stroller, set him loose in the playroom, took him to the auditorium for a movie or a performance by jugglers or clowns. The nurses watched with pride as Gregory learned to hold his bottle, drink from a cup, feed himself, and pull himself up on the rails of his crib.

Greg spoke his first word, to a nurse: "Mama."

He took his first steps.

More weeks passed, and it was Greg's first birthday. The nurses brought him a cake with one candle. Regulations forbade them from lighting it, but that was okay—the nurses had presents for their birthday boy. They blew up balloons and they sang, and everyone drank sweet cherry punch.

It was February, and Greg's parents hadn't been able to return; Frances was pregnant again, for the twelfth time.

It was February, and Lillehei was deeply frustrated.

By the second week of January, Lillehei had wearied of noodling around on dogs. Gregory Glidden had improved markedly, prompting cardiologist Ray Anderson to declare: "Patient now looks as good as at any other time that I have seen him. Probably as ready for surgery as he will be anytime." Lillehei told Wangensteen that he was about to take cross-circulation into the operating room.

And Wangensteen said no.

Once again, a critical point for Lillehei coincided with a turbulent period in his mentor's personal life. It seemed as if Wangensteen had lost his first son—his namesake—forever.

Arrested in late 1951 for stealing and forging checks, Bud had been jailed and brought to court, where he explained to a judge how he, the twenty-two-year-old son of a distinguished professor, had turned into a petty thief. Bud said that he had dropped out of college, failed at a number of menial jobs, and squandered the money his mother and father gave him. With two young children to support, Bud said, he had to do something.

"I was very desperate for money," said Bud. "I had been telling my wife that I was working."

Bud explained how he had discussed his plight with a friend who was in similar circumstances.

"We were talking together," Bud said, "and we were discussing possible ways that we might obtain money—first, entirely honest ways. Then, as a process of elimination, we finally came down to burglarizing and stealing checks and passing them."

"Sort of a business," the judge said.

"Well, yes, as a money-making proposition."

Bud explained how he had read two books on locksmithing,

then used a needle and a bottle opener to break into a warehouse, where he and the friend stole the checks.

Bud pleaded guilty to forgery in the amount of $97.73 and was sent to the state reformatory for men four days before Christmas in 1951. Less than two months later, Bud's twenty-one-year-old wife filed for divorce—claiming, among other things, that Bud had once tried to kill her with a hammer, that he had hit her when he was drunk, and that while she was still in the hospital after the birth of their second-born son, Bud had slept with another woman. The court granted the divorce and gave Bud's ex-wife custody of the children.

Bud's father, meanwhile, had concluded that his son was insane.

No one could deny that Bud was unusually gifted. Not only did he have a genius IQ, but he was a talented musician, a charming conversationalist, and even a marvelous dancer—none of which Professor Wangensteen was. Bud had an evident if offbeat talent in writing and someday, he believed, he would be a celebrated author.

"I'm going to make the Wangensteen name more famous than it is now," Bud told the woman who became his second wife.

But Wangensteen believed his son needed to be institutionalized, not published. Bud was no sooner released from the state reformatory when Wangensteen arranged to send him to a state mental hospital, where he was confined for seven months. Wangensteen then sent him to a private institution.

He was released before 1953 ended. But the rift between the son and the father was by then unbridgeable; before the decade was over, Bud left for Spain, never to return.

Wangensteen was faring no better with his wife; their long-rancorous marriage had reached the point of dissolution. The Wangensteens bickered constantly—often over Bud, to whom, unlike the other two children, Helen had always been close. Helen's drunkenness and depression had worsened: on at least one occasion, she had tried to kill herself with an over-dose of sleeping pills.

Helen had had enough of the distinguished Professor Wan-gensteen. She wanted a husband who made more money, and a father who was more sympathetic to their gifted but way-ward son. In 1953, Helen threw Owen out of the house; he took a room at a St. Paul hotel. Then she also threw out their other son, a twenty-year-old medical student, and arranged to have the wheels taken off his car so he could not drive it. The Wangensteens' daughter this time escaped her mother's wrath, for she was married and no longer lived at home.

In January of 1954, Helen filed for divorce.

Through his lawyer, Wangensteen blamed his wife for Bud's troubles: "Such difficulties as have occurred between the par-ties hereto have largely occurred because of [Helen's] desire to coddle her said son and not require him to realistically face life and accept its responsibilities."

In early 1954, the celebrated chief of surgery seemed to have but two things to sustain him. One was a woman he'd re-cently met and was beginning to fall in love with—Sally Davidson, the lovely, refined, and wealthy editor of a medical magazine.

The other was his beloved surgeons.

And now, two of his best were at odds.

Accomplished at repairing ASDs under hypothermia, John Lewis by early 1954 had set his sights on the next milestone, VSD. Despite his friend Walt Lillehei's doubts that cold was the answer to that more difficult defect, Lewis believed he could succeed. Moreover, he outranked his friend: he had per-formed history's first successful open heart operation, and he was Lillehei's senior on the faculty by a year. Lewis asked Wangensteen's permission to be the first in Minnesota to at-tempt correction of a VSD, and Wangensteen granted it.

111

When Lillehei learned this, in the second week of January, he went to the chief immediately.

Let John try a few first, Wangensteen told Lillehei. Then it will be your turn.

Lillehei was incensed—as incensed as he ever got. "I hesitate to be persistent to the point of obnoxious," he wrote to his boss after their meeting, "however, sometimes one is driven on by the conviction that he may be right."

Concerned that Wangensteen might think that he was motivated only by the desire to go down in history, Lillehei noted that Charles Bailey had just published a report of closing a VSD using hypothermia. Bailey's report of the operation, on a fifteen-year-old girl in March of 1953, did not indicate whether she was still alive, however, nor whether the closure was complete—whether the case was successful. In any event, Bailey had not attempted another VSD, nor had anyone else undertaken even one.

"My dear Walt," Wangensteen wrote to Lillehei, "your persistence surprises me. . . . I did not say that you should not do your case now. I merely indicated that I thought it would be the nice thing to do, in light of Dr. Lewis having pointed specifically toward repair of interventricular defect, to let him do the first few cases here. . . . Friendly, helpful, interested and sympathetic competition is good."

Gregory Glidden would have to wait.

And Gregory did, a pawn in a game of ego. He waited several more weeks in Variety Club Heart Hospital while Lewis, with his hypothermia, operated on a child with a suspected VSD.

But the diagnosis was wrong. The child had a more severe defect, which Lewis could not fix, and the child died on the table.

Lewis brought a second child with a suspected VSD to his operating room, and this time, the diagnosis was correct.

But that child, too, perished on Lewis's table.

Lewis was devastated: losing anyone tore him up, but a kid especially. He'd go home with a horrific headache and crawl off to bed, allowing only the interruption of his wife bringing hot packs for his pounding temples.

Maybe Walt Lillehei was right: maybe hypothermia wasn't the answer, after all. In any event, Lewis had lost his passion for pioneering open heart surgery. Like John Gibbon, he was dropping out of the quest.

It was Walt's turn now. God be with him.

LILLEHEI WITH (*CLOCKWISE*) WARDEN, VARCO, AND COHEN

8

Brave Hearts

C ASUALLY CONSIDERED, the heart is merely a pump.
In fact, it is a sophisticated organ. Comprised of four
valves, two holding chambers, two pumping chambers, a spi-
derweb of vessels to nourish it, and an intricate network of
nerves to control it, the heart must harmoniously propel blood
through approximately sixty thousand miles of arteries, capillaries, and veins—two thousand gallons of blood through the
average-size person each and every day.

It does not pump autonomously. The heart must properly
respond to sometimes severe, sometimes subtle changes in
temperature, exertion, emotion, and stress. It should last a
normal lifetime: some two and a half billion beats over the
course of seventy-five years. Evolution has confirmed the
heart's importance by protecting it with bone. The brain and
bone marrow, foundation of the immune system, are among
the few other parts of the body so honored.

Congenital disease can warp the heart with great variety.
Valves can be sealed tight, missing parts—or absent altogether. Major vessels can be misplaced, narrowed, or blocked
completely. A chamber can be too small or missing, a wall too

thick or thin. The heart's electrical system—its nerves—may go haywire. The muscle can be weak. Holes may occur almost anywhere, in almost any size. Studying heart pathology, one is reminded that the genetic symphony that produces a normal baby is indeed a wondrous and delicate one.

Heart abnormalities weighed on Walt Lillehei's mind as he counted down the days to Gregory Glidden's operation. He examined freshly autopsied hearts and formalin-preserved hearts in the University of Minnesota's modest collection, and he combed texts and illustrations. His consuming interest in March of 1954 was of course the VSD.

Because no one had been able to cure one, no one had ever taken much interest in the defect. Little had been written or depicted, and the rare drawings and photographs that Lillehei unearthed gave a view only through the left ventricle. Lillehei would be entering Gregory's heart through the right ventricle, which was easier for a surgeon to access than the left.

And if a VSD were simply a basic hole, Lillehei might have taken a measure of comfort. But he had learned enough to know that a VSD could be large or small, in plain view or carefully hidden, in an area of thick wall or thin. It could actually be several holes, and it might have more sinister features as yet unknown. Catheterization was useless to divine any of this, and holes bored in dog hearts in the attic of Millard Hall revealed only so much.

And all this was assuming the diagnosis was correct.

Until he'd opened Gregory's heart—perhaps even after—Lillehei would be confronting an enigma.

If anyone could help prepare him, it was Jesse E. Edwards.

A pathologist at the Mayo Clinic, Edwards had become fascinated with anatomy while in medical school. War sent him to Europe, where his autopsies of Allied soldiers helped convict war criminals at the Nuremberg trials. No part of the body bored Edwards, but he became infatuated with the heart and the study of its many diseases. Edwards, too, had joined the quest.

He was certainly at the right place for a pathologist. Unlike other medical centers, the Mayo had saved almost every heart from every autopsy since early in the century; by 1954, it undoubtedly had the largest collection of human hearts anywhere. What a sight to behold! Untold thousands of hearts, preserved in formalin inside wooden pickle barrels—infants' hearts, old people's hearts, normal hearts, and hearts laid waste by every conceivable disorder and disease.

More remarkable still was Edwards's connection to his specimens, which approached the surreal. Encased in a cheesecloth sack, each heart was tagged, but Edwards could reach into a barrel, fish virtually any one out, and—prompted only by the anatomy—recite the deceased person's age, sex, history, and doctor (though not always the person's name). Edwards could expound on each heart's defect and, if known, the cause. Further details that he had not memorized could be gleaned from the X rays, electrocardiograms, and the many other records that he kept for each of his thousands of hearts. Edwards was like a medieval monk, preserving sacred knowledge.

One day in the middle of March 1954, only a week before he was to operate on Gregory Glidden, Lillehei drove south to Rochester in his 1951 Buick Roadmaster convertible. Lillehei took Warden and Cohen, who'd helped develop cross-circulation, and Varco, the senior surgeon who would be his primary assistant in the operation on Gregory.

Jesse Edwards was an altruistic man; he believed his collection belonged to science, not to him or to the Mayo. He told Lillehei and his colleagues to look at whatever they wanted, to ask questions, to take their time.

They spent the better part of a day. Edwards had almost fifty hearts with VSDs, and the surgeons explored every one.

By nightfall, they needed a drink.

Lillehei's old haunt Mitch's was gone, but the Parker House had proved a worthy successor. It was owned by Red Dougherty, a jazz pianist whose Dixieland band had been one of Walt's favorites at Mitch's. Red still performed. He drank hard liquor and played piano in his new club while his guests dined on Kansas City steaks and drank martinis in the smoky semidarkness.

When Walt and the boys walked through the door that March evening on their way home from the Mayo Clinic, Dougherty lined them all up with drinks. Walt was more than one of Red's regulars: not long before, he had successfully operated on Dougherty for constrictive pericarditis, a disease in which the heart's outer lining becomes inflamed and scarred, causing pain with each breath. Dougherty could never thank Lillehei enough.

Over drinks, the surgeons dissected what they'd seen in Rochester. They were discouraged.

Lillehei had known VSD was a clever adversary, but until today he hadn't appreciated just how cunning it could be. Even if Gregory Glidden's VSD diagnosis was correct—a courageous assumption—they were less certain than ever about what awaited them inside the child. They had examined nearly fifty VSDs in Edward's collection, and no two were exactly the same.

And that was but part of their worry.

Whether they found a VSD or some previously unseen defect, just getting inside the heart would be perilous. Lillehei's reliance on the azygos factor would reduce bleeding, not eliminate it; if the defect was hidden, would they be able to find it amid the blood? What about the invisible nerves that crossed the area where Lillehei would be sewing—one misplaced stitch would wreak havoc! And who could say that the stitches would hold? In healthy dog hearts, the stitches had held, but the lower chambers of the sick human heart were different. The force of pumping might rip the stitches out of tissue already weakened by disease.

And what about the second patient in this strange new operation—the donor?

One drink at Red's led to another, then a third, and pretty soon Walt and the boys were feeling all right. They hadn't come this far to be put off by pickled hearts.

Next week, they would take cross-circulation clinical.

Late in the afternoon of March 25, 1954, the supervisor of the University Hospital operating rooms, Genevieve A. Scholtes, typed the schedule for the next day. She mimeographed copies and, as usual, had them distributed throughout the hospital.

A quick glance suggested that March 26 would be just another day. Wangensteen was operating on a woman for cancer, and Varco was down for partial removal of a stomach. Walt Lillehei's first case would be a hernia repair, the sort of surgery that helped pay the bills.

But Lillehei and Varco were also listed for "repair of IV septal defect" on patient number 862270, Gregory Glidden. Beneath the little boy's name, someone had penciled in "Lyman Glidden" and "cross circulation."

Scholtes knew what *this* was about; Lillehei had invited her to the dog lab a few days earlier to determine how to configure the operating room for this untested, two-person surgery. Wangensteen and his inner circle also knew, of course, but no one else.

Still, such a curious item could not long escape notice in a teaching hospital, where gossip is sport. By early evening, the schedule had come to the attention of the chief of medicine, Cecil Watson.

An expert in liver disease, Dr. Watson had been one of the leading medical consultants for the Manhattan Project, the effort that produced the atomic bomb. The public knew him as a distinguished internist and professor—and avowed hater of pigeons. "Apart from the above mentioned dangers of disease transmission from droppings, dirt or dust," he wrote in one of his many letters to the editor on the subject, "the nuisance occasioned is not inconsiderable. The constant billing and cooing is like the Chinese water torture, gentle at first but steadily less easy to endure."

Watson and Wangensteen respected each other, but they sometimes seriously disagreed. The two chiefs had locked horns over ulcers, Watson being of the serve-them-milk-and-cream school, Wangensteen of the cut-the-stomach school. Watson had once pulled a patient headed for the operating room off the elevator before Wangensteen could get to him.

When it came to cardiac disease, Watson wanted every patient admitted first to a medical ward, where his people could dictate the treatment—including deciding if surgery was warranted—while Wangensteen wanted potential surgery candidates admitted straight to the surgical service, where he was boss. This conflict had complicated planning of the Variety Club Heart Hospital in the late 1940s, when Watson had demanded control of all of the hospital's beds, and Wangensteen had insisted on at least half. Wangensteen won that battle, although not the chief of medicine's affection.

Now, reading the OR schedule for March 26, 1954, Watson quickly figured out what Wangensteen's protégé planned to do.

Once more, the chief of medicine was rankled.

Of all of Wangensteen's surgeons, Walt Lillehei was unquestionably closest in spirit to crazy Charlie Bailey. Lillehei's brilliance was beyond dispute, but his wartime experiences seemed to have drained him of fear—and if Rommel and the beach at Anzio had not, then his brush with lymphosarcoma surely had. And it was hardly a secret that as far as Wangensteen was concerned, Lillehei all but walked on water.

This time, Wangensteen and Lillehei had gone too far.

But Watson himself could not stop them. So he went to the

only one who could, Ray M. Amberg, the director of University Hospital.

———

Amberg was an administrator, but also an effective politician and flack. At budget time every year, he went to the Minnesota legislature to make University Hospital's case. He knew how the system worked—knew the importance of a cigar and a drink with a committee chairman, of free lunches and free tickets to University of Minnesota games. Amberg made sure that his hospital's doctors, Wangensteen and Lillehei included, understood that it was their solemn duty never to charge when a legislator or his relative needed care.

"I used to think it was a one-horse outfit till I injured my arm in an elevator and went over there," one member of the House Appropriations Committee said at a hearing in 1953. "I went through that place and now I don't care what they ask for—they're going to get it if it's in reason."

Amberg selectively courted the press, allowing reporters and photographers into University Hospital to trumpet the staff's many marvelous successes. Such triumphs as John Lewis's with hypothermia were better than free tonsilectomies come state budget time.

On this March 25, Watson outlined for Amberg his objections to Lillehei's proposed cross-circulation operation.

Had Wangensteen and his men so soon forgotten John Lewis, the hospital's celebrated open heart pioneer, who thought he could fix a VSD—and couldn't? Had the lessons of Dennis, Gibbon, and everyone else with a heart-lung machine been lost? And cross-circulation, said Watson, was more menacing than any machine. With a machine, you might lose one person, but with cross-circulation, you could lose two, one in the bloom of health. Lillehei might achieve a first, all right: for the first time in history, he might have an operation with a 200 percent mortality. What kind of ethics were those? And just picture the headlines if the press got hold of the story. Watson believed that Lillehei, encouraged by Wangensteen, was nearing the point of madness.

Wangensteen defended his permission to Lillehei. There was no denying that two patients would be put at risk, but the risk should not be exaggerated; cross-circulation had succeeded spectacularly in the lab. As for Lillehei—well, Walt might be fearless, but he wasn't reckless. He believed he could save Gregory Glidden. Without Walt, the baby would die.

Amberg had helped adjudicate Wangensteen and Watson's Variety Club Heart Hospital dispute, but this time, he wasn't stepping in. These heart surgery cases were strange: Lewis had brought University Hospital glory, and thus far, they'd managed to keep the deaths out of the papers.

Amberg had a hunch about Lillehei. If Wangensteen had faith in him, then Amberg would, too.

———

The Gliddens had been waiting. Lyman and Frances visited their son in early March, but work and family had otherwise kept them in Hibbing. Just keeping the household afloat was a daunting business. Lyman worked overtime in the mines, but with periodic layoffs and payments on a new refrigerator, television set, and car, money was scarce. To make ends meet, the Gliddens grew their own vegetables, and they kept chickens, a goat, and a cow. Lyman hunted deer for venison, and Frances canned fruit.

That March was particularly tight: another of the Gliddens' children was sick, with a terrible case of eczema, and the medical bills for him and for Gregory were mounting. On top of everything, Frances was eight months pregnant.

On March 17, Dr. Lillehei had called to say Gregory's operation could finally be scheduled. Lyman would be his son's donor; they shared the O-positive blood type.

And so, a week later, Frances and Lyman made the two-hundred-mile drive to Minneapolis. They found their son in fine spirits. Gregory was laughing and entertaining himself in a playpen, and his temperature was normal, his appetite hearty. Except for the telltale murmur in his little chest, he seemed a healthy, if skinny, little boy.

It was the evening of March 25, the fifth day of spring.

Over in the main hospital, Amberg, Watson, and Wangensteen were debating cross-circulation.

On the third floor of the Variety Club Heart Hospital, Gregory was finishing his dinner. Because his parents were visiting, he was allowed to stay up until seven-thirty, an hour past his regular bedtime. The Gliddens kissed him good night and went to Lyman's room, a floor below their son's.

Gregory drifted off easily. He was still sleeping soundly at six the next morning, when a nurse came to prepare him for surgery.

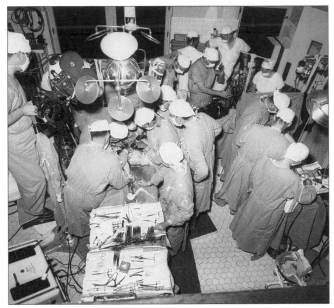

CROSS-CIRCULATION IN THE OPERATING ROOM

9

Zero Time

A FTER A SOUND night's sleep, Walt Lillehei awoke at the usual hour, 6 A.M. He ate breakfast and read the morning paper.

Atoms dominated the news in March of 1954. Albert Einstein, whose genius had unlocked the secrets of matter, had just celebrated his seventy-fifth birthday. The federal government was funding development of nuclear-power plants, which promised electricity in unlimited quantity at discount prices.

But the biggest headlines evoked horror, not hope. President Eisenhower had announced he was prepared to meet "Communist aggression" in Russia and China with "massive instant retaliation." All month, stories had been leaking out of Washington about the secret test of a U.S. hydrogen bomb whose destructive power dwarfed anything before it. Detonated on March 1 at Bikini, a Pacific atoll, the weapon exploded with fifteen megatons of force, more than a thousand times stronger than the bomb that obliterated Hiroshima. Castle Bravo, as the weapon was code-named, had left a crater in solid rock 250 feet deep and more than a mile wide—and unleashed a fireball nearly four miles in diameter. "It looked to me like a diseased brain up in the sky," one observer recalled. Eighty-two nautical miles away, snow-white ash drifted down on the crew of the regrettably named *Lucky Dragon,* a Japanese tuna boat, causing radiation burns. "Some of the *Lucky Dragon* catch had to be pulled off the market and buried as likely to cause tumors if eaten," reported the Associated Press.

The irony could not have escaped Lillehei.

When he thought about it, infrequently now, he wondered if radiation had been a factor in saving his life. Nearly four years after surgery and X-ray therapy, Lillehei's lymphosarcoma had not recurred. Lillehei felt and looked great; the only sign of his cancer was the faded scar on his neck. He did not have cataracts, one of radiation therapy's possible side effects. He was full of energy, an impassioned young man who regularly worked eighteen-hour days.

Lillehei drove to University Hospital, changed into scrubs, dropped by Operating Room II, then went to another room and started his hernia operation, which would take about an hour.

Over in the Variety Club Heart Hospital, the nurse had awakened Gregory Glidden. She washed him and dressed him in a fresh gown, and a resident gave him a preoperative sedative. An orderly appeared; after soothing the baby, he wheeled Gregory across a bridge into the main hospital.

In Room II, the anesthesiologists were waiting.

———

Room II was small, although larger than the rooms to either side. Two tables fit inside it, barely. It had a white-tile floor, green walls, steam radiators, and windows, which on hot summer days were opened in hopes of catching a breeze from the Mississippi River.

This did not look like a place where history could be made,

but it already was. Here, three years before, Clarence Dennis had been the first to use a heart-lung machine on a person. Here, Patty Anderson had perished in a river of blood.

The equipment in Room II was state of the art for 1954, but that wasn't saying much. Doctors and nurses monitored vital signs the traditional way—with glass thermometer, blood pressure cuff, and fingers against a wrist. The surgeon could use an electrocardiogram machine, but electrical interference from suction pumps and motors made it unreliable. Nothing was computerized; nothing was digital. A surgeon had no pulse oximeters to continuously monitor the oxygen content of the blood, no pacemakers or defibrillators to rescue a heart in trouble, no speedy blood-gas analyses to warn of unseen complications. One of the most useful instruments for Lillehei's purposes was the wall clock, which had a large second hand.

Although tremendous strides had been made since the days when a patient was put to sleep with an ether-soaked rag, anesthesia remained hazardous. Precise monitoring of metabolism was impossible with such limited instrumentation. Sophisticated ventilators did not exist, so doctors administered anesthetic gases by squeezing a black rubber bag—fifteen or twenty squeezes a minute, minute after minute, hour after hour—for as long as the surgery continued; maintaining a steady, even flow by hand was difficult. Sick babies were particularly at risk under anesthesia because of their size and fragility.

And though the highly inflammable ether had been largely abandoned, other explosive agents had taken its place. Cyclopropane would be used on Gregory. Even the name smacked of danger—it sounded like an experimental jet fuel. Indeed, cyclopropane was dangerous stuff: a spark from a faulty outlet or switch could set it off, as could static electricity. Anyone entering an operating room was supposed to wear special shoes, and women were asked to forego silk underwear, which, rubbing against a uniform, could generate deadly sparks.

127

Despite the precautions, cyclopropane still blew up with unsettling frequency. A prominent Harvard professor estimated that as many as 130 explosions occurred annually in American operating rooms, and about thirty were fatal. No one knew the true count: in 1954, doctors could still bury their mistakes.

Death by explosion was ugly. It blew out the anesthetic circuit, sending lethal glass and metal shrapnel flying—and destroying the patient's lungs in a scorching, bloody burst. "The explosion lifted the mask about an inch or more from the face," one doctor reported of a death he'd witnessed. The worst accidents took out a doctor or a nurse, too.

"I don't understand how something like this should happen in an up-to-date hospital," the grieving husband of a woman killed in a cyclopropane explosion in Brooklyn told *The New York Times*. A doctor was knocked unconscious by the blast, likely caused by static electricity. Such a spark, the hospital superintendent lamented, could have come from anywhere, "by moving an arm or a leg, or even while sitting in a chair."

It was the worst sort of peril: a deadly phantom, like the defect that awaited Walt Lillehei inside Gregory Glidden's heart.

———

Gregory was put to sleep. Finished with his earlier case, Lillehei came back into Room II with Varco.

It was eight-thirty in the morning.

Varco was a demanding, sometimes gruff man who scared the more timid of the younger residents. Short and beefy, with a shock of thick black hair, he had big paws for hands—meeting him on the street, you might have guessed he worked the waterfront. But in fact he was the slickest of the University of Minnesota surgeons, and that included the chief. "When Dr. Varco gets through with a case," a nurse who knew his work remarked, "it looks like it's the way the Lord made it."

Varco began closed heart operations in the early 1940s,

when Wangensteen lost interest. More experienced than anyone in Minneapolis, Varco was now one of the best cardiac surgeons in America—and, despite his temperament, he was a
great teacher, as Lillehei and many others could attest. Varco
introduced several new closed heart techniques and modifications, and he was pleased to lend his expertise to Dennis and
Lewis. But sometime in the early 1950s, Varco lost his enthusiasm for the quest to operate inside the open heart. Some
speculated that oxygenators and cross-circulation hookups
mystified him, although that seemed unlikely; Varco was too
smart. Some believed he'd once lost a little girl who reminded
him of his daughter, and that was too close to home.

Or maybe losing Patty Anderson—and then Sheryl
Judge—had distressed him. Maybe he had no desire to be
Columbus without a compass, navigating rivers of blood;
maybe, like Lewis and Gibbon, the death of innocents undid
him. So many had died already—how many precisely, no one
would ever know. No government agency or professional association kept statistics, and while some surgeons reported their
failures in the medical literature, many did not, for there was
no glory in dead kids. It was easier to quietly bury them and
move on.

Whatever his reasons (and he never said why; confession
was not his style), Varco was content to play second fiddle
on these open heart cases. And Lillehei was pleased to have
him.

All set? Lillehei said.

The people with him were.

Lillehei cut into Gregory.

He doesn't have any subcutaneous tissue, thought Lillehei.
Not an ounce of fat. His heart will be so tiny.

Assisted by Varco, Lillehei cut through Gregory's sternum
to expose the boy's heart. So far, no surprises.

It looks okay, Lillehei said. You may bring in the father
now.

It was 8:45 in the morning.

Groggy from sedation, Lyman Glidden was brought into Room II and positioned on a table close to Gregory. Lyman could almost have reached out and touched his son.

Technically, a local anesthetic would have been adequate to blunt Lyman's pain this morning. But the chance an anxious parent would pop up off the table at some critical moment to check on his baby was not one Lillehei could afford to take. The anesthesiologists put Lyman under, but they trod lightly: anything in Lyman's bloodstream would soon be in his son, who, less than a tenth the father's size, could die of an overdose before anyone realized what had happened.

Working together, Cohen and Warden cut into Lyman's right groin, exposing two major vessels: the femoral artery, which carries freshly oxygenated blood direct from the heart, and the saphenous vein, which returns blue (oxygen depleted) blood to the heart. Cannulas were inserted into the vessels and connected to the beer hose, which went to the T-6S milk pump. There was no cooling blanket; hypothermia would not be used.

Warden and Cohen had been in Room II since before dawn preparing the pump. Calibration was a weak link: too little blood into Gregory, and his oxygen-starved brain would waste away; too much blood, and the pressure would cause general tissue swelling, also potentially fatal. The donor-patient exchange had to be exactly equal: the amount of blood that left Lyman and went into Gregory had to be exactly the amount that left Gregory and came back into Lyman.

Lillehei and Varco, meanwhile, attended to their half of the circuit, Gregory.

Using a stiletto-bladed knife, Lillehei made a nick in a vein in the baby's neck and pushed a cannula down into the superior and inferior venae cavae—the large vessels that channel depleted blood into the heart. Lillehei passed another cannula through a large artery into the aorta, which sends reoxygenated blood to the body, and connected the cannulas to separate lengths of beer hose, which led through the pump to Lyman. Then Lillehei siphoned the air out of the circuit.

Gregory's heart still pumped his blood through his body, but soon his blood would be detoured into his father.

It was 10:03 A.M.

Lillehei was a study in intensity. He wasn't thinking of making history, or of the two open heart tragedies this room had already seen, or of the consequences of failure. He was thinking only of this surgery. This was another of his talents: his ability to blot out everything but the task at hand. Athletes in a later era called it the zone. Lillehei was deep in the zone.

Unlike sports, in which performers have the benefit of practice, radical new surgery does not allow the luxury of a rehearsal. Lillehei was relying on experiments with dogs, and on a detailed tactical plan, devised in discussions with the anesthesiologists, the OR supervisor Genevieve Scholtes, and his surgeons. He was relying on luck and skill.

Lillehei examined Gregory's half of the circuit—the tubing and cannulas inside the boy—and was satisfied. Warden and Cohen had prepared Gregory's father, and both sets of anesthesiologists were satisfied with the anesthesia. A technician manned the pump. Nurses bustled. Observers peered down from the balcony, and on the operating room floor, residents and interns perched on stools.

Ready? said Lillehei.

Everyone was.

Pump on, Lillehei said.

The technician hit the switch and the T-6S started up. In his operative record, Lillehei declared this "zero time."

Lillehei watched blood flow through the beer hose for two minutes, long enough to confirm that the circuit did not leak, and that the donor-patient exchange was equal. Then he tied small tourniquets around Gregory's venae cavae and his pulmonary artery, which sends blood to the lungs.

Gregory's heart was now detoured; twenty-seven seconds later, Lillehei cut into its wall. Varco gently parted the edges of the incision as Lillehei lengthened it with a pair of scissors. In less than a minute, the heart was open.

———

It continued to beat, as expected.

But even with the reduced rate of flow determined by the azygos factor, the blood into the right ventricle was more than Lillehei had anticipated. Not as bad as with a machine but bad enough. Thank God, Varco was able to keep ahead of it with the suction; thank God, they could see. They did not have the fabled dry field—a completely bloodless, still heart—but they could see! They had a freshet, not an angry river.

Lillehei began to explore the inside of Gregory's heart with his fingers and forceps. He did not have magnifying lenses, nor proper light—his headlamp threw too narrow a beam. But beggars can't be choosers: cardiac headlamps didn't exist in March of 1954, and Lillehei had borrowed his from an otolaryngologist, who used it to look inside ears.

Still, Lillehei found the defect easily. It was exactly as catheterization had foretold: a simple VSD, a single hole the size of a dime. It was high on the septum, the wall that divides the heart's chambers, making it readily accessible.

The diagnosis seems to be correct, Lillehei said. Let's go ahead.

Lillehei began sewing. He was deliberate, quiet—fast but not frenzied. A dozen ordinary stitches, and the VSD was shut.

Any seamstress could have sewn that up in five minutes, Lillehei had thought holding Dorothy Eustice's lifeless heart.

In the balcony, silence.

On the stools, stillness.

It was twelve minutes and fifteen seconds past zero time.

Gregory's heart had been open for almost ten minutes, well beyond hypothermia's accepted outer limit. His skin was pink.

Lillehei inspected his work. It was good. The stitches were holding; the tissue was solid and the closure was tight. If the stitches held, nature would permanently seal Lillehei's handiwork in a matter of days.

Careful not to trap air, Lillehei closed the outer wall of the heart with a silk running stitch. Then he looked for signs of heart block—damage caused to the electrical conduction system by a stitch piercing an invisible vital nerve. Heart block was a poorly understood condition, but Lillehei knew it could be fatal. The heart could slow, then stop altogether, and no one could prevent it.

But Gregory's heart beat vigorously. Lillehei put his hand on it. The vibration was gone now—the murmur, too.

It looks good, Lillehei said. Varco agreed.

Someone in the balcony snapped a picture.

Release the cavae, Lillehei said.

Varco loosened the tourniquets that had closed the vessels and Gregory's heart took over. They were fifteen minutes and twenty seconds past zero time.

Lillehei waited three more minutes to be sure Gregory's heart continued to beat normally. It did. The boy's blood pressure and pulse were acceptable. His skin was still pink.

Pump off, Lillehei said.

The T-6S, designed to move milk, completed its human debut.

They were nineteen minutes past zero time. Anesthesia had been gloriously uneventful. Lillehei and Varco disconnected

Gregory from his father and closed his chest, while Warden and Cohen unhooked Lyman and sewed closed the incision in his groin.

Lyman was coming to. Is Greg okay? he soon asked.

Warden said he was.

Gloves still wet with blood, Lillehei and Varco reached across Gregory and shook hands.

———

As Lillehei's dogs had done, Gregory awoke promptly and showed no evidence of neurological damage.

I'm very pleased, Lillehei told Lyman. The defect was what we expected, and we got a good closure.

Lillehei went to the waiting room to tell Frances that her husband and son were fine. Frances was so grateful. It was the answer to a prayer.

After five hours in the recovery room, Gregory was returned to Variety Club Heart Hospital. There were no intensive care units in 1954, but unusual cases warranted private nurses. One was assigned to Gregory around the clock.

Gregory slept most of the rest of the day. Occasionally he opened his eyes, but Lillehei had prescribed Demerol for pain and it quickly pulled him back under. The boy's vital signs remained stable, his color was good, his breathing occasionally labored but not alarmingly so. He urinated, an indication that his kidneys had made it through the surgery undamaged.

Lillehei knew all this firsthand because he checked on Gregory repeatedly: in person that afternoon and evening, by phone from home at three-thirty the next morning, in person again at dawn. And when Lillehei wasn't there, Varco or one of Lillehei's trusted residents was. Lillehei could have been a first-time father with his newborn boy.

Gregory's progress on the first full post-operative day, March 27, encouraged Lillehei. Offered sugar water, Gregory drank "eagerly," one of his nurses noted; an hour later, he took milk. Although still under the influence of Demerol, he was "more active now, kicking legs and moving arms." He went on to spend a good night. When Lillehei called at 5:45 the following morning, Gregory was sound asleep.

The little boy continued to improve on March 28. That morning, he had his first meal since before his operation: a bite of poached egg and three bites of Cream of Wheat. By afternoon he wanted chocolate milk, his favorite. His stitches itched and he tried to scratch them. He sat up.

"Breathing sounds clear now," a nurse wrote.

Frances and Lyman visited Greg one more time and then headed back north to Hibbing, hopeful that Lillehei had spared their baby their daughter's fate.

By the first of April, Gregory seemed to have reached a turning point.

His appetite had returned with a vengeance—he was eating heartily and drinking orange juice and chocolate milk in reassuring quantities. His cheeks were that lovely pink, and he slept well. Gregory had become University Hospital's star patient, attracting crowds of curious students and doctors.

Yet there were alarming developments. Gregory's breathing was becoming labored, and his lips sometimes turned blue. His left arm seemed cold, his right arm hot. He kept tugging at his ear, as if it hurt. Once so cuddly, he flinched now when anyone got near—and he cried when he was held. He wanted to be alone, in the hissing seclusion of an oxygen tent.

"Seems less alert," noted one nurse on April 2.

"Definite asthmatic type wheezing," noted another nurse, the next day.

Suspecting a respiratory infection, to which heart patients are prone (when their lungs have been damaged by high blood pressure), Lillehei prescribed antibiotics.

April 3 was a Saturday. Lyman had the day off from the mines, and he drove down from Hibbing with Gregory's older sister Geraldine. The boy was in his oxygen tent when they got to Room 310, at about two o'clock in the afternoon. Gregory did not recognize his father or his sister. He refused to be held. All he wanted was a sip of chocolate milk.

Lyman and Geraldine went out into the hall.

Suddenly, a nurse ran from Gregory's room to the nurses' station. Almost immediately, Varco and two other doctors arrived. Behind the closed door to Room 310, Lyman and Geraldine heard a terrible commotion.

Gregory was suffocating—his breathing passages had swollen and he couldn't get enough air. The doctors placed an oxygen tube down into his trachea, but Gregory continued to suffocate, leaving Lillehei no choice but to perform an emergency tracheotomy. Gregory was rushed to surgery.

When the orderly brought Gregory back two hours later, he seemed better. He drank his chocolate milk, his color returned, and he breathed easily again. Except during periodic suctioning through his tracheotomy tube, he slept through the night.

———

The Sunday paper that April 4 teased of summer. Lawn seed and roses were advertised for the first time since last year, and the Home section published its annual planting guide. Easter was almost here and the weather had turned delightfully warm. The long prairie winter was finally over.

"Color good. Moving about very well alone," a nurse wrote of Gregory.

Believing that Gregory had finally turned the corner, his father and sister set out for home. They left Gregory quietly fondling his beloved blankie.

But that night, the boy wheezed continuously, twice becoming dangerously short of breath, and the next day he was lethargic and weak. The nurses got a bit of pudding into him but he wasn't interested in chocolate milk. By 6 P.M., Lillehei had him under constant observation.

Lillehei himself was in and out all evening—with Varco, Warden, Cohen, and others, a batallion of doctors against an elusive enemy. Once, Gregory stopped breathing altogether; Lillehei saved him by massaging his tiny chest until his lungs restarted. Suctioning was bringing up blood and yellow mucus. A pink froth smeared Gregory's lips.

After midnight, a nurse placed a telephone call to Hibbing. Geraldine answered the phone.

Your brother's taken a turn for the worse, said the nurse. Your parents had better get down here.

But, once again, Gregory rallied.

Exhausted, Lillehei went home to sleep. He could be back, if needed, in a matter of minutes.

———

As the night wore on, Gregory drifted in and out of consciousness. He took a sip of sugar water at 3:30 A.M., but thereafter could not be roused; not even the constant suctioning of his tracheotomy tube woke him. A nurse found his blood pressure and pulse only after a frighteningly long hunt.

By dawn, an anesthesiologist was squeezing life into Gregory with a rubber bag.

Lillehei arrived at eight. Gregory suddenly opened his eyes, looked around, then withdrew into himself again.

Lillehei listened to the boy's heart and examined his chest. His blood pressure and pulse were fading rapidly. His skin was the color of rain clouds and he was cold to the touch.

At nine, Gregory opened his eyes again, as if taking one last look.

Then he closed them.

Ten minutes later, his heart stopped.

But Lillehei would not surrender.

He told the anesthesiologist to keep bag-breathing the boy while he readied a shot of epinephrine, a powerful stimulant.

Aiming between Gregory's ribs, Lillehei plunged the needle directly into the little boy's heart. Then he waited.

And waited.

Nothing.

Lips like rain clouds.

Stillness.

Cold.

Gregory was gone. His heart, so badly burdened for so long, could not be coaxed back.

At 9:15 A.M. on April 6, 1954, Lillehei pronounced his first open heart patient dead.

———

Lyman and Frances Glidden, due to deliver her twelfth child that very week, arrived in Minneapolis shortly before noon to find that Gregory had already expired. Frances and Lyman signed for their son's T-shirt and pants, and then they went to see Lillehei.

I'm sorry, Lillehei said. We gave him our best try.

The Gliddens said they understood. They had known the risk going in.

Lillehei believed that pneumonia or some other complication, not the surgery, had killed Gregory, but there was only one way to know for sure. Would the Gliddens consent to an autopsy? The results would help Lillehei determine if he would try cross-circulation again in the operating room, in the hope of saving other patients, or retreat to the lab.

Lyman Glidden signed the permission slip. Before leaving, he and Frances thanked Lillehei and wished him luck in the future. Then they went home with their grief.

———

It was two o'clock in the afternoon when the pathologist started on Gregory. He proceeded through the abdominal organs and finally got to the lungs and heart, which he cut out and handed to Lillehei. Two summers before, in this room he now knew so well, Lillehei had said farewell to Dorothy Eustice.

Lillehei opened Gregory's heart for the second and final time.

He was not deeply religious. But he was praying that in the eleven days since zero time, Gregory's heart had healed.

PAMELA SCHMIDT WITH HER PARENTS, APRIL 1954

10

Lourdes in Minneapolis

B ABY PAMELA SCHMIDT clung to life so precariously that one of her doctors once suggested that her parents waste no time in having another child. But somehow, Pamela survived. She was four years old when she entered Variety Club Heart Hospital for her latest evaluation, on April 3, 1954.

The Schmidts knew about Gregory Glidden, whose room was near Pamela's, and they soon heard of Dr. Lillehei, the daring young surgeon who had just fixed Gregory's heart in a spectacular new operation. Pamela had the same problem as Gregory—a VSD.

Gregory's death seemed like one more cruel blow to the Schmidts. And yet, Dr. Lillehei was curiously encouraging when he talked to them after the little boy had gone home in a hearse. Surgery hadn't killed Gregory, said Lillehei; a fast-moving case of pneumonia had. Lillehei wanted to try his new open heart operation again, and the cardiologists had identified Pamela as a candidate.

Lillehei described cross-circulation and explained the risks to both patient and donor. Then he told Ronald and Mary Schmidt to go home to their small apartment in Minneapolis

141

and think it over. Pamela was sick, but not quite at death's door.

What emotions over the next few days! One minute, the Schmidts wanted to rush back and give Lillehei their permission . . . and the next, they thought of poor Gregory, cold inside a coffin.

In the end, the Schmidts decided they had no choice. With or without surgery, their own little girl was sliding toward the grave.

In fact, even as Ronald and Mary gave Lillehei their permission, Pamela was battling yet another case of pneumonia. The girl's temperature rose to fever level, her appetite disappeared, and she awoke scared and crying in the middle of the night. "General malaise," a nurse wrote in her record.

Yet again, the Schmidts' hopes seemed dashed, for Lillehei could not operate while Pamela was so sick.

But the girl improved with penicillin, and Lillehei sent the Schmidts to the hospital blood bank to see if either parent's blood type matched Pamela's.

Once more, the Schmidts were bitterly disappointed: Ronald's blood type indeed matched Pamela's, but his blood was low in hemoglobin and the blood bank director recommended that he not serve as his daughter's cross-circulation donor. But Lillehei believed the director was being overly cautious. Against the hematologist's advice, the surgeon scheduled the operation.

And so, on the morning of April 23, 1954, an orderly wheeled Pamela Schmidt into University Hospital's Room II. Lillehei and Varco opened the girl's chest while Warden and Cohen prepared the father. Lillehei examined the outside of Pamela's heart and saw nothing unusual—no nasty surprises. The surgeons connected the patient to the donor.

Pump on, said Lillehei.

It was 11:05 in the morning, zero time for Pamela Schmidt.

The pump started and a father's blood mingled with his daughter's—the parent's life supporting the child's.

Lillehei cut into the wall of Pamela's heart, widened the opening, and began to explore the interior of the heart. The interior was almost completely dry; by modifying his technique slightly (another tourniquet, around the aorta), Lillehei had managed to stem nearly all of the bleeding.

Lillehei quickly found Pamela's defect: a gaping VSD, about the size of a half-dollar. The diagnosis was correct—no nasty surprises inside the heart, either.

Sewing as he had done inside Gregory Glidden's heart, Lillehei closed the hole with six stitches. He tested the closure and found a leak, which he sealed with a seventh knot of silk. So far, the operation was blessedly smooth.

Lillehei was ready to wrap things up when trouble suddenly developed: the chambers of Pamela's heart were contracting too slowly, and they were out of synchronization. Lillehei feared that he had accidentally caused heart block: damage to the heart's electrical system. He considered removing the seventh stitch—it might have pierced an invisible nerve—but he could not risk a leak. The seventh stitch had to stay.

Gambling that Pamela's heart would regain proper rhythm on its own, Lillehei sewed up the heart wall. But the chambers continued to beat slowly and erratically, and now, Lillehei was certain that he'd caused heart block. He feared that the heart would keep slowing and finally stop altogether.

It was ten minutes and forty-five seconds past zero time.

They were quietly frantic, for no one in the spring of 1954 knew how to reverse heart block.

The seconds passed, and then a life was spared: Pamela's heart began to beat faster, its chambers working again in smooth coordination. Lillehei observed the heart for two more minutes, until he was satisfied it could continue on its own.

Pump off, he said.

It was thirteen and a half minutes past zero time.

Pamela Schmidt was cured.

Gloves still glistening, Lillehei and Varco reached across their patient and shook hands.

A day or two later, Lillehei got a call from Victor Cohn, a reporter at the *Minneapolis Morning Tribune.*

Lillehei had never spoken to Cohn, but he knew his work, which was impeccable, and often on the front page. Young and enterprising, Cohn had established a medical beat at a time when most newspapers ran only wire-service medical stories. Cohn often showed up at scientific meetings, and he spoke the language of doctors. He maintained excellent contacts inside the University of Minnesota Medical School.

I understand you have a fantastic new operation, said Cohn. Something that's never been done.

What have you heard? said Lillehei.

Cohn said that he already had enough for a credible account. But he needed Lillehei's cooperation for an even better story— one, he suspected, that the wires would send around the world. It seemed as if Lillehei had achieved victory in the open heart quest.

We're not ready to announce this yet, Lillehei said.

Cohn pressed: If Lillehei wouldn't talk, the reporter would run with what he had.

Let me decide what we want to do and I'll get back to you, said Lillehei.

As informed as Cohn was, he knew nothing of the events inside the autopsy room the afternoon of April 6, the day Gregory Glidden died. The pathologist had handed Lillehei the boy's heart and Lillehei had reopened it.

What beauty within! Lillehei's stitches had held—the hole had been fixed! This proved unquestionably that pneumonia, not cross-circulation, had killed his first open heart patient.

Elated though Lillehei was, he understood that Gregory's death could only be fresh ammunition for the detractors. Lillehei could hear them now: *So you closed a VSD. The result was no different than Clarence Dennis's. You didn't kill two patients, but you did kill one. How many more must die?* As word of Gregory's death spread through University Hospital—eventually reaching the internists, including the chief of medicine, Cecil Watson—pressure would surely intensify to restrain the chief of surgery's protégé.

Lillehei left the autopsy room and went directly to his cardiologist friends, who agreed that despite Gregory's death, cross-circulation deserved another chance.

Don't find another patient, Lillehei instructed the cardiologists; find two. We'll schedule them immediately, one right after the other. Before anyone can blink, we'll prove this thing works.

The cardiologists quickly identified two candidates: four-year-old Pamela and a three-year-old boy, Bradley Mehrman, who also had a VSD.

Once again, Lillehei sought his boss's approval. Wangensteen had congratulated him after his operation on Gregory, and Lillehei doubted that news of the boy's death would make the chief of surgery flinch, given the results of the autopsy. Still, in light of John Lewis's tragic experiences with VSDs, Lillehei sent Wangensteen a note outlining his intention to operate on two more children.

"The parents want it done, even though we informed them of the eventual outcome in the first case," wrote Lillehei.

"Dear Walt," Wangensteen wrote back. "By all means, go ahead!"

In the end, Lillehei decided that no single reporter deserved to break the news on cross-circulation. So the University of Minnesota News Service called a press conference. Victor Cohn felt that he'd been cheated.

Nevertheless, Cohn was present the afternoon of April 30, when Lillehei took the stage in the Variety Club Heart Hospital auditorium, where not so long ago jugglers and clowns

145

had entertained Gregory Glidden. Several other reporters also attended, and Lillehei had brought his surgeons.

Lillehei had never hosted a press conference—no surgeons in 1954 did—but his presentations at Broadmoor and many subsequent scientific meetings had given him polish. He spoke of the sad conclusion to the first attempt at cross-circulation, and the two wonderful successes that had followed, just in the last ten days—a boy and a girl, ages three and four. He showed slides: a diagram of the operating room setup, a drawing of a VSD, and a photograph of Room II on the morning of the operation on the little girl. He thanked everyone involved in the effort, including the nurses, the anesthesiologists, and the blood bank director. He noted, proudly, that a $500 pump that could be purchased through the mail was the technological core of his breakthrough.

"We long have felt that there must be some simple method of working inside the heart," said Lillehei. "When elaborate machines designed as a substitute for the heart and lungs proved unsatisfactory, we tried using the animal's own lungs to purify his blood, and substituted the simple mechanical pump for his heart."

That idea, Lillehei said, had logically led to cross-circulation. "Our method," said Lillehei, "is widely applicable by surgeons experienced in heart surgery."

A news service official distributed a four-page press release, and it could have ended there.

But this was Walt Lillehei, a man who drove a Buick convertible and wore gold jewelry. Walt believed in celebrating success. Walt knew what reporters wanted. The chief of medicine had feared the headlines if cross-circulation had been a disaster . . . well, here was something for that curmudgeon Watson.

The door opened and a four-year-old girl in a yellow dress appeared. With her brown eyes and dark bangs, Pamela Schmidt was impossibly cute. She sat in a wheelchair (for safety), but she looked in the bloom of young health—rosy cheeks and a bewitching glimmer of a smile. It seemed incredible that only one week before, she had been on the operating table, her heart wide open—her heart for a moment seemingly about to stop for good—and that before her appointment with Lillehei, she had been an invalid, slowly slipping away.

Pamela posed for photographs with her parents and Dr. Lillehei. Her father, a factory maintenance worker, answered Victor Cohn's questions.

Was it scary, heading into Room II as his daughter's donor?

"No," said Ronald, "I didn't have any worries—none at all. My wife—she's the one that does the worrying."

Schmidt talked about how very sick their daughter had been: eight or nine bouts of pneumonia in her young life, most of her first year spent in an oxygen tent. How the Gliddens and Patty Anderson's parents would have understood all that.

"She's a little fighter," Schmidt said. "She's never given up."

From now on, predicted her surgeon, Pamela would be normal. And she was—as was the three-year-old, Bradley Mehrman, whose surgery had actually been three days before Pamela's.

It was a fairy tale. It was Lourdes in Minneapolis. It was a welcome antidote to talk of Communism and mushroom clouds in the shape of diseased brains.

The story went worldwide, on the radio and in print. *Time* magazine called cross-circulation "daring." *The New York Times* proclaimed: IMPOSSIBLE SURGERY NOW DONE. A paper in California decreed Lillehei's work a "miracle," while one in Egypt called it "revolutionary." Said the London *Daily Mirror:* "It was an operation as extravagant and fantastic as any ever written in a shilling science 'thriller.' " The only detractor was an antivivisectionist, who sent Lillehei a clipping of the *Mirror* story with a scribbled salutation: "YOU SWINE. God did not cre-

ate his animals to be tortured by man. Experiment on your-
selves and convicts."

Not surprisingly, the hometown paper provided the most
extensive coverage. The story in the May 1 *Morning Tribune*
was bannered atop the front page and was chock-full of photos, including an exclusive shot of Bradley Mehrman and his
mother, and another purporting to show Drs. Lillehei and
Varco operating on Pamela. In truth, as a reader might have
guessed noticing that neither surgeon's mask covered his nose,
it was a staged shot, taken by the University News Service—
in the autopsy room, of all places! Well, no need to split hairs.
That photo, too, was published worldwide, as dramatic illus-
tration of The Operation.

Newspapers had written about Lillehei before: the *Tribune*
had mentioned a grant he'd won, and the *Evening Star,* the *Tribune*'s sister publication, had noted the national award for his
study of failure in dog hearts. But those were filler. The publicity surrounding cross-circulation blew into a blizzard, and
Lillehei was beseiged with inquiries from reporters, doctors,
and worry-sick parents of dying children. Lillehei could not
turn anyone away; he himself had almost met death.

Nor did he forget the Gliddens. On May 4, amid all the
publicity and plans to push ahead with cross-circulation,
Lillehei found time to send a letter to Frances and Lyman.

"It is still a source of bitter disappointment to me," he
wrote, "that we were not able to bring Gregory through the
post-operative period after the operation had seemingly gone
so well. I do wish to tell you again that had it not been for the
encouragement gained from Gregory's operation, we would
not have had the courage to go ahead with these other children. I feel greatly indebted to both of you."

Lillehei received a reply from Frances. So much had happened in a month. Gregory had been laid to rest in a tiny
white coffin next to LaDonnah—in an unmarked grave, for
the Gliddens could not afford another stone. And just hours

after burying one son, Frances had delivered another. So far, please God, the Gliddens' twelfth child seemed healthy.

"Though it is so hard not to feel bitter that little Greg couldn't have lived to rejoice with the other two," Frances wrote, "we just have to accept it as the Lord's will and we know his death wasn't in vain as it has given these two children another chance to live and no doubt many more. . . . May God bless and guide you in the wonderful work you are doing."

———

The ink had barely dried on the first news stories when Lillehei and his surgeons traveled to Montreal's Sheraton-Mount Royal Hotel, where the American Association for Thoracic Surgery was meeting. Much of the Broadmoor crowd would be there. What a four years it had been. John Lewis's success with hypothermia had inspired countless surgeons, and now, it seemed, no major university lacked some kind of open heart program, or plans to soon start one.

Still, only Lillehei could claim success with ventricular septal defects. And only Lillehei seriously intended to soon attempt repair of yet more extreme defects—the biggest monsters, such as tetralogy of Fallot.

Surgeons had packed the Mount Royal auditorium when Herb Warden stood to present the Lillehei group's paper, which detailed cross-circulation in the laboratory dog. When Warden exceeded the allotted ten minutes, the warning light on the podium began to flash. The moderator, who was supposed to cut long-winded types off, ignored it; he was enraptured, as was the audience.

Because the paper had been submitted before the surgeons had used cross-circulation in the operating room, it fell to Lillehei during the discussion period to talk about the three children. What a figure he cut! How self-assured!

Lillehei outlined their experiences with Gregory Glidden

and Pamela Schmidt, then described their difficulty with little Bradley Mehrman: how the surgeons had needed twenty-four minutes—*twenty-four minutes!*—of open heart time to fix his defect, and yet the child had still come out of surgery completely healed. Anticipating criticism of the potential for losing two patients in one operation, Lillehei said: "None of the donors to date has suffered any untoward effects. They have all been discharged from the hospital 24 to 36 hours post-perfusion. . . ."

When he finished, Lillehei was mobbed by surgeons who were suddenly considering cross-circulation themselves. Clarence Dennis, still unsuccessful with his heart-lung machine in New York, even said to Richard Varco: Gee, Dick, do you think I should continue work on this oxygenator business?

But not everyone extended congratulations.

It might be simpler and certainly cheaper just to make another baby, remarked one surgeon with a black sense of humor.

Robert Gross offered no praise; he still smarted from the previous year's meeting of the association, when Lillehei had publicly criticized Gross's atrial well, in which a surgeon worked with his hands submerged in a pool of blood. "Surgical procedures carried out under direct vision are far more likely to be satisfactory than those carried out blindly," Lillehei had said. In the wake of that blunt critique, Gross had fired off a nasty letter to Wangensteen. How dare he let his brash young surgeon criticize the great professor's work?

Predictably, John Gibbon also dismissed cross-circulation. Gibbon already considered Lillehei a twerp—and a regular party boy at these conventions, carrying on until all hours in the hotel rooms and bars like a sailor on leave. And now this flood of news stories with Lillehei posing for the camera like a home-run king. Gibbon himself had refused photographs when *Time* had published its story about his one success ("too camera-shy," the magazine noted). And to think Lillehei had called a press conference, too, before presenting his results to the medical community. The temerity!

And for what? For Lillehei's latest cockamamy idea, successor to that silly azygos factor notion of his.

"We are still convinced that it is preferable to perform operations . . . by some procedure which does not involve another healthy person," said Gibbon in Montreal. "There must be some risk to the donor in a cross-circulation."

In fact, the greater risk was the same as with a heart-lung machine, the risk to the patient; while cross-circulation gave a surgeon more time inside the open heart, those extra minutes created unprecedented problems. Surgeons would have to invent new operating techniques, then refine them; they had to develop new equipment, such as suitable headlamps. The skills of a Florence Nightingale no longer sufficed to nurse the patient back to health after surgery—managing the post-operative effects of chronic heart failure on other organs, notably already damaged lungs, remained largely virgin territory.

Lillehei succeeded with his first two cases after Montreal, but the next one ended in failure on the operating table: expecting a VSD, Lillehei found an atrioventricular canal, the defect that had crippled Patty Anderson. Lillehei attempted to repair it anyway, and mistakenly sewed the aortic valve partially shut. The patient died.

But he pushed on, curing six of his next eight patients—the two deaths were from post-operative complications, as Gregory Glidden's death had been. Although surgeons elsewhere had begun laboratory investigation of cross-circulation, none had yet used the technique on a person—and still no one had developed a safe heart-lung machine. During the spring and summer of 1954, Lillehei was the only person in the world performing advanced open heart surgery.

151

Then came terrible trouble.

On September 7, 1954, a seven-month-old girl died of heart block four hours after Lillehei operated. Lillehei's next patient, two weeks later, died an hour after the chest was closed. The patient after that—dead. A week later, one more. From September through the middle of November, Lillehei used cross-circulation on seven patients—and only one lived.

The doubters once again were alarmed. Certain cardiologists at Lillehei's own hospital refused to refer him patients, lest they die on his table. Behind Lillehei's back, nurses had started calling him "murderer."

That autumn, Lillehei was deeply shaken. Now, he, too, very nearly recoiled from death.

It wasn't death alone: it was telling parents that he'd lost their precious baby. Unlike some senior surgeons, who assigned the job to residents, Lillehei always broke the news himself. He owed parents at least that. He wanted them to know that their sacrifice was not wasted, that he'd done the very best he could.

And it was the frustration of failing with the highest possible stakes, a human life, something he struggled to put into a philosophical context.

To the residents and anyone else who questioned why the path was so tortuous, Lillehei would say, You don't venture into a wilderness expecting to find a paved road.

He would say, Good judgment comes from experience, and experience comes from bad judgment.

And he would remember Dorothy Eustice, a woman too young to die.

Yet there were still moments when he wondered what level of hell he'd dragged them all down into. To what end—to build Jesse Edwards's collection of pickled hearts?

Then he would go home, have martinis and a steak dinner with Kaye after the children were all tucked safely in their beds, but he would not talk about how his losses affected him. He would sleep, and the next morning he and his surgeons would scrutinize the latest results. They would try something new in the lab, use a different approach in the operating room, find fortification in their successes and Lillehei's renewed conviction that some day they really would get it right—and children as yet unborn would be saved. He would recall Werner Forssmann, the seemingly mad German who'd invented heart catheterization, and others like him who'd ventured into the darkness when everyone else had kept to the coziness of a well-lit room.

Nature gives up her secrets with great reluctance, Lillehei believed.

But Nature in the autumn of 1954 was not only reluctant. She was stubborn and cruel, and drenched in blood.

Dispirited though he sometimes was, Lillehei tried to keep his faith that complex open heart surgery was here for good when he traveled, near year's end, to Atlantic City for his profession's most prestigious conference. The American College of Surgeons was meeting, and most of the University of Minnesota's surgical staff was attending. Still strongly supportive of Lillehei, despite that autumn's troubles, Owen Wangensteen especially looked forward to Atlantic City. Owen had washed his hands of his son Bud by now, and a judge had granted his wife, Helen, the divorce she wanted. Owen that summer had married his true love, Sally Davidson—a woman who not only loved him in return, but deeply admired his work, regardless of what it paid. They had just returned from a Hawaiian honeymoon.

Lillehei in Atlantic City spoke again about cross-circulation. The publicity surrounding the first three cases had subsided, but Lillehei had two more sensational feats to report. Just two weeks ago, having learned from his first failed effort, he became the first person in history to correct an atrioventricular canal, a worse defect than VSD.

More astonishing, on August 31, he'd been the first to cure a child with tetralogy of Fallot.

Named for Etienne-Louis Arthur Fallot, the nineteenth-century Frenchman who first described it, tetralogy was a nightmarish cardiac cluster comprised of a VSD, a narrowing of the pulmonary valve, an excessively muscular right ventricle, and a mispositioned aorta. Victims' distinctive blue cast was the result of deoxygenated blood being routed back into the body before it could be replenished in the lungs.

To surgeons in 1954, tetralogy was the Mount Everest of heart defects.

Alfred Blalock had improved the prognosis with his blue-baby operation. Many who'd seen tetralogy of Fallot in the autopsy room doubted it could ever be completely corrected, even with breathtaking technique and a way into the open heart that gave ample time and a relatively dry operating field. Some in Atlantic City were openly skeptical of Lillehei's claim, a skepticism he countered by inviting doubters to come watch him work.

On the alert for the next big heart surgery scoop, reporters covering the Atlantic City meeting cared less for the arcana of tortured anatomy than the unprecedented circumstances of Lillehei's historic August 31 operation. For the first time in fourteen cross-circulation cases, Lillehei had found no ready donor for his patient, Mike Shaw, a ten-year-old who lived in central Minnesota. A total stranger had saved the day.

Mike Shaw had been sick almost from birth. Taking just a few steps turned his lips and fingertips blue and left him gasping for breath; lacking a car, his mother pulled the tyke around town in a little red wagon. In the spring of 1954, Mike fell critically ill. His doctors scheduled a blue-baby operation at University Hospital—but then, only a week after curing Pamela Schmidt, Dr. Lillehei got wind of the boy. Lillehei judged Mike a candidate for cross-circulation and he gave a choice to Mike's parents, who by then were divorced: the blue-baby operation, or Lillehei's own new open heart surgery. But it had to be one or the other, for if the blue-baby procedure was performed, Lillehei could not then justify a second operation on the same sick heart.

The Shaws entrusted their son to Lillehei, who arranged for blood testing. Mike had the rarest of all blood types, AB-negative, which occurs in just 1 percent of the population—but neither of his parents did. Relatives, neighbors, and friends were tested, but none of them matched, either. The Red Cross and a veterans' organization combed their files and found Howard Holtz, a twenty-nine-year-old highway worker.

Married and the father of three boys, including a baby, Holtz did not know the Shaws, but he had read about Dr. Lillehei and his celebrated patient Pamela Schmidt. He drove to Minneapolis to hear Lillehei explain the donor's role in cross-circulation.

It's the only chance we have to save the boy's life, said the surgeon.

Although an appendectomy was the only surgery he'd ever undergone, Holtz agreed then and there to be Mike's donor. "I just wanted to do what I hope someone would do if my child were a blue baby," he later told the Associated Press.

Shortly before the operation, Holtz met Mike, his three brothers, and his mother, who worked in a poultry-processing plant cutting the legs and wings off chickens.

He's sure a nice little guy, said Holtz to Mrs. Shaw. I hope things go well.

And they did—for both the heroic cross-circulation donor and the ten-year-old patient. "The following day," wrote Mrs. Shaw in the diary she kept of her son's disease, "we noticed the color returning to his lips and fingernails and ears. He was turning a nice rosy pink."

———

Lillehei's colleagues in Atlantic City had reached no consensus on cross-circulation. At least one other surgeon had now tried the technique on a person, without success; some surgeons remained enticed, while others had turned disdainful. Hearing of Lillehei's many losses, more than one doctor had decided Lillehei was unethical; morphine for the pain was better for a dying, heart-crippled child than surgery that seemed an immediate death sentence, asserted one internist. Cardiologist Helen Taussig, of blue-baby-operation fame, also condemned Lillehei. Learning of his one success with tetralogy of Fallot, Taussig said, Too bad, now he'll continue.

Many doctors figured it best to wait until a heart-lung machine was perfected, if one ever was. They shared Gibbon's concern about the potential for 200 percent mortality. True, Lillehei had not lost a donor, but rumor had it that a young woman had sustained severe brain damage.

Lillehei was finishing his Atlantic City presentation when someone from the back of the audience called out: Admit that you have a vegetable in the hospital!

The heckler did not know her name, but he was referring to Geraldine Thompson.

Like most cross-circulation patients in Minneapolis that autumn, eight-year-old Leslie Thompson, born with a VSD, had traveled a long, sad road before reaching Lillehei. She'd visited general practitioners and heart specialists in Massachusetts, Colorado, and Texas, where a cardiac group led by the pioneering surgeons Michael E. DeBakey and Denton Cooley had seen but not cured her. Leslie had been X-rayed and catheterized, poked and probed—and until Lillehei, all the doctors could promise was an early death.

Counting down the days until Lillehei operated on their daughter, Geraldine and Dan Thompson fretted. Perhaps a round of golf would ease their minds. They were at a country club when an older man overheard them talking.

I've been a practicing surgeon in this town for over thirty years, said the man. I know Lillehei and Varco and Warden and the whole crew over there. They're very optimistic, but you're crazy if you do this. It's just too new.

The Thompsons were startled. *Crazy*—why, that was the very word rattling around in the back of their heads, behind their desperation to save their daughter.

Geraldine and Dan contacted Lillehei immediately. Lillehei said the decision was still theirs, of course, but he didn't think they should be put off by some stranger's idle comments. A physician friend of Dan's agreed. And after agonizing all over again, the Thompsons decided to proceed with cross-circulation. Geraldine said she would do anything to save her daughter's life; her blood type matched Leslie's, and she would be the cross-circulation donor.

On the morning of October 5, 1954, Dan kissed Leslie and she was taken away to the operating room. Then he went to Geraldine's room and kissed her, too.

I'll do my best, Geraldine said.

Dan retired to the waiting room. Dan's physician friend would be watching the operation from the balcony.

It seemed to Dan that hours had passed when his friend appeared—but it was really only minutes. The friend looked pale. The news was not good.

Something went wrong, said the friend. They've stopped the operation.

Lillehei had closed Leslie's chest before even getting to her heart. But the problem wasn't with Leslie. It was with Geraldine.

Lillehei found Dan Thompson a short while later.

We made a horrible mistake, said the surgeon. Somehow Geraldine was subjected to an air embolism.

Bubbles in her blood.

Like Sheryl Judge, Clarence Dennis's second open heart patient, Geraldine had been given air. The mistake was entirely that of an anesthesiologist—a doctor who wasn't even on Lillehei's team (a floating supervisor, he was trying to be helpful). But Lillehei didn't get into those details, not then. He apologized to Mr. Thompson, and he held out faint hope: air had poisoned Geraldine's brain, but some neurological function remained. How much, only time would tell.

Geraldine Thompson left the operating room in a coma. She lingered in the darkness for several days and then she slowly emerged. But she was not the same. She dragged her left foot when she walked and she clenched the fingers of one hand, as if clinging desperately to something no one else could see. Time and place had become a confused jumble, and she sometimes hallucinated—saw spiders and snakes in her bed. She could not speak properly anymore. She could not care for herself, let alone four small children—one with a heart that remained unhealed.

———

Pamela Schmidt, meanwhile, had been crowned. Quite literally, she had become the Queen of Hearts.

Plucked from obscurity by virtue of a hole in the heart—a hole that any seamstress could have sewed up in five minutes—young Pamela appeared on network TV, and Hubert Humphrey and other U.S. senators sent greetings. The very month that the Thompsons traveled to Minneapolis with their daughter Leslie, *Cosmopolitan* magazine published a six-page spread on the Schmidts. In the months and years to come, Pamela would be state, then national Queen of Hearts for the American Heart Association. With their constant stories, the Minneapolis newspapers would make Pamela perhaps the best-known child in Minnesota in the 1950s. They would photograph Pamela riding her bike, playing in the snow, watching herself on TV, listening to her mother read her "Squiffy the Skunk," appearing with Dr. Lillehei at a Heart Association fund-raiser.

Lillehei had been transformed, too, and not only publicly.

Wangensteen would soon promote him to full professor.

Despite his detractors, Lillehei was in demand for medical school committees, though he wasn't much impressed; committees, he believed, were largely forums for the exchange of hot air, and so he rarely attended their meetings. Lillehei did not aspire to be a chief or, dear God, a dean.

Indeed, anything that smacked of bureaucracy put Lillehei off. A model of precise discipline in the operating room and in the lab, Lillehei kept a slovenly office, with journals, records, and X rays reaching toward the ceiling. Lillehei always procrastinated answering letters and returning telephone calls, except from patients. Bills for his services were late going out; incoming checks were dumped in a desk drawer, where they stayed uncashed for months. Walt even missed the deadlines for his income tax filings—Uncle Sam would have to wait, just like everyone else.

But there was relief, for Lillehei had discovered that, if ignored, most paperwork went away as the piles grew and grew.

In any event, Walt had work. Plenty of real work, and not licking stamps.

Lillehei didn't need the accident with Geraldine Thompson to realize that the concerns of John Gibbon and Cecil Watson, while exaggerated, were not irrational; the need for a donor during cross-circulation did endanger a second person.

Cross-circulation had other shortcomings. Because it intimidated many surgeons and internists, it was unlikely ever to fulfill Lillehei's grand ambition, which was making complex open heart surgery available to all. Even if they so desired, doctors at an ordinary community hospital, far from the front lines of medicine, could probably never convince their director or the trustees to support such an unconventional program.

And physiological limitations—the extra load on the donor's lungs and heart—made cross-circulation more suit-

159

able for fixing the hearts of sick children than sick adults. Of the forty-five cases Lillehei would attempt, the heaviest patient weighed eighty-three pounds. Most weighed considerably less.

As winter set in and Lillehei continued with cross-circulation in the operating room, he and his residents explored an abundance of new concepts in the lab.

One idea seemed the height of simplicity: the so-called arterial reservoir, in which surgeons dripped donated blood from a bottle into a patient whose heart was clamped off. Lillehei used the reservoir successfully, but only five times; like cross-circulation, it best suited children.

Another idea arose with that time-honored favorite, the laboratory dog. The idea had much in common with what a Canadian doctor was doing with monkeys. What the Canadian was doing could give an anti-vivisectionist an ulcer.

Like Wilfred G. Bigelow, whose hypothermia research had inspired Lewis to perform the world's first successful open heart operation, William T. Mustard worked in Toronto, at the famed Hospital for Sick Children—right across the street from Bigelow's hospital. Trained as an orthopedist, the young Bill Mustard could not resist the heart's allure. Bones just couldn't compare, for Dr. Mustard was an adventuresome sort—a man who, in the middle of a formal dinner or speech, would suddenly demonstrate his talent for one-armed push-ups. And if that didn't get a laugh, Mustard might swallow a live goldfish or dive into a fountain, tuxedo and all.

Cold did not tempt Mustard. Like others, he believed that a surgeon could best get inside the living heart by detouring the blood around it, and he understood that the great challenge was not in the pumping, but in the reoxygenating of depleted blood. While others experimented with steel screens and disks—with artificial lungs—Mustard investigated the real thing: lungs from human cadavers. Getting nowhere with them, Mustard then turned his attention to live rhesus monkeys, which were available and cheap.

In 1951, Mustard decided he was ready to operate on a person.

He first chose a sixteen-month-old child who was suffering from tetralogy of Fallot. As his assistants put the child to sleep, Mustard anesthetized four monkeys, shaved their chests, cut out their lungs, and flushed the lungs clean of monkey blood with an antibiotic and saline solution.

"Except for the scar tissue around parasitic lesions," Mustard wrote, "the lungs eventually become wholly white in appearance." Mustard then suspended the monkey lungs inside bell jars, into which pure oxygen was forced, and linked the lungs with tubing that connected to a pump. After priming the circuit with human blood, Mustard hooked up the patient.

But the surgeon had trouble making the final connection to that first patient, and the patient died on the table.

Mustard's next patient, a ten-month-old blue baby on whom he operated a year later, also died, two hours after surgery.

But Mustard kept going. By 1954, he had brought a dozen children into his operating room, one as young as nineteen days, none older than eleven years, and all close to death. He did not get a single survivor.

A dozen children—and 100 percent mortality. Most died on the table or shortly thereafter, although one, a baby boy, lasted fifteen days.

———

Gilbert S. Campbell did not know of Mustard's monkey-lung work when he, too, embarked upon the quest. Campbell was another of Owen Wangensteen's resplendent young surgeons—at the age of thirty, he had already published nineteen scientific papers.

Why not canine lungs for reoxygenation? Campbell wondered.

He began experimenting, using dogs as both lung donors and patients, and by the end of the winter of 1954 to 1955, he was almost ready to try his dog-lung technique on a human. But first he needed to determine if the lungs of man's best friend could refresh a man's blood.

And so one day, during a noncardiac operation inside a man's groin, Campbell (with the man's prior permission) brought tubing, a pump, and two freshly cleansed dog lungs into the operating room. He connected one end of the lungs to one of the anesthetized man's veins, and the other end to an artery. Blue blood from the man flowed into the dog lungs— and bright red blood flowed out, then was returned to the man. The man showed no ill effects.

Campbell was ready for an open heart case. Because Lillehei and Varco were more experienced in cardiac operations, he enlisted them as the lead surgeons. Walt was always eager to try a new idea—his own or someone else's.

The first attempt, to fix a boy with tetralogy of Fallot, ended with the patient's death shortly after surgery.

A few days later, Calvin Richmond arrived on the scene.

A thirteen-year-old African-American from rural Arkansas, Calvin had been riding on a horse-drawn ice truck in August of 1954 when he fell, was run over by the truck, and was knocked unconscious. He awoke in the hospital, where he spent the next nine days. But Calvin never fully recovered: he had trouble breathing and he could not shake a stubborn cough. Finally, doctors at the University of Arkansas Hospital diagnosed a ventricular septal defect, caused when the wheel ran over his chest. No Arkansas surgeon could fix a VSD, but they'd heard about Lillehei.

Son of a sharecropper, Calvin was too poor to get to Minnesota on his own, so a newspaper and a TV station in Little Rock took up his cause. People contributed nearly $3,000, much of it pennies donated by schoolchildren, and the

Arkansas Air National Guard provided a plane for the nearly thousand-mile voyage. A reporter flew to Minneapolis with Calvin and Calvin's mother on March 16, 1955. "I wonder if he recognizes what preparations have already been made for him ahead," wrote the reporter while they were circling to land, "how many hours the surgical team who will perform the operation has already spent in the study of his individual case—much less the years of training and practice they've put in to develop the skill that makes them dare to gamble with life." Gamble indeed. No one had raised the specter of attaching the boy to a mongrel dog's lungs.

And, in fact, Lillehei planned to use cross-circulation—and to operate for free. But Calvin's mother, whose blood type matched her son's, refused to be put under; she'd never heard of Geraldine Thompson, but instinct told her she might never wake up. She'd grown up in the Deep South; she had her convictions.

Lillehei then turned to the state penitentiary, where he had previously found a donor for a cross-circulation case in which none of the patient's relatives' blood matched.

But these inmates were white, and none would let a black man's blood mingle with his—not even the blood of a poor sick boy.

So Lillehei decided to go with a dog lung.

Calvin was a strapping lad, and on the morning of his operation, Campbell chose the largest animal they had: a country dog, its lungs purer than a city hound's. Campbell sedated the dog with sodium pentothal, then killed it with a second, stronger dose. He quickly harvested the dog's lungs, placed them in a sterile basin, flushed them clean, and carried them from his lab to the operating room, where they were suspended in a plastic chamber and suffused with oxygen. Beer hose connected the lungs to Calvin, and that old standby, the T-6S pump, moved the boy's blood through the circuit.

For twenty minutes, the lungs of a stray dog kept young Calvin alive while Lillehei sewed up the hole in his heart.

A day later, Ray Amberg let a reporter visit Calvin in the Variety Club Heart Hospital.

"Whatcha doing?" the reporter overheard Calvin say to a nurse who was putting a rubber cuff around the boy's arm.

"I'm just taking your blood pressure—to see if you're alive," said the nurse.

"You don't need to do that," said Calvin. "I'm still talking, ain't I?"

———

Calvin Richmond left for home a month after his operation. Lillehei deployed the dog lung in another twelve cases, but, like Cohen and Warden's old self-lung, it was finicky: it had a tendency to bloat and become useless. Several children died before the Minnesota surgeons finally gave up on it.

Dog lung, self-lung, arterial reservoir, cross-circulation— none was ideal for open heart surgery. Lillehei was leading the quest, but there remained a distance to cover before true victory could be declared. Miracle headlines notwithstanding, Lillehei now believed that he needed a *machine*.

Safe, simple, easily duplicated, obviously much different from Dennis's or Gibbon's . . . but still a machine. If only someone had the design.

Certainly no one would have placed money on the man who'd shown up at University Hospital one day shortly before Lillehei operated on Gregory Glidden. Richard DeWall wasn't even a surgeon. He practiced general medicine in a Minneapolis suburb, and he was bored. In his spare time, at his kitchen table, he'd built a plaster-of-Paris model of an artificial heart valve.

DeWall wasn't looking to change history when he met Dr. Lillehei. He only wanted to know what a big-time University of Minnesota surgeon thought of his little creation.

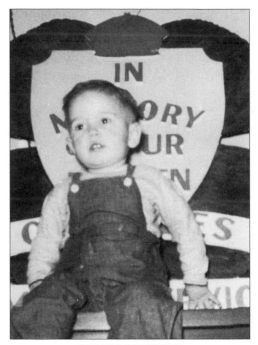

JIMMY ROBICHAUD

11

Needles and Bubbles

W HEN HE ARRIVED at University Hospital with his
plaster-of-Paris heart valve model, Richard DeWall
called first on Richard Varco, whom he remembered as one of
his teachers in medical school. Varco was impressed. Only
dreamers in 1954 believed artificial heart valves would ever
become reality, but DeWall was evidently ingenious. Who
knew what other ideas he might come up with?

Varco sent DeWall to see Lillehei, who found it difficult to
refuse anyone with ambition and desire. Still, the young man
seemed an unusually poor prospect. DeWall was soft-spoken,
shy—a suburban family doctor, familiar with a stethoscope
but barely acquainted with the scalpel. DeWall had interned
at an undistinguished hospital, and his University of Minnesota Medical School grades were nothing to be proud of. In
fact, he had graduated near the very bottom of his class.

DeWall explained that his interest in the heart went back to
the closing days of the war, when, as a navy draftee, he'd con-
tracted rheumatic fever, which can damage the heart's valves.
Since then, DeWall said, life as a general practitioner hadn't
proved to his liking. Research would be more exciting.

167

DeWall's grades did not dissuade Lillehei—he himself had nearly flunked high school chemistry, prompting his teacher to predict he'd drop out of college—but Lillehei knew that the dean of the University of Minnesota graduate school, who had final say in admitting surgical residents, would reject DeWall for his lackluster academic performance. In any event, the department of surgery didn't have the money for another residency. The dean believed the chief of surgery already had too many residents running around.

We're working on a shoestring budget, said Lillehei. I'm sorry, Dr. DeWall, but we just don't have a position for you.

But Lillehei did have the money to hire DeWall as a part-time animal attendant in his lab, and the family practitioner, just twenty-seven-years old, jumped at the offer.

DeWall fit right in. Cheerfully, he tended the dogs, mopped the floors, calibrated and ran the laboratory T-6S pump, and, before long, assisted in canine operations. He proved extraordinarily clever, with an affinity for mechanical devices and an innate understanding of synthetic materials and how they interacted with living cells. Lillehei knew he'd uncovered a treasure. DeWall closed his family practice and went to work full-time in Lillehei's lab.

Weeks passed. Lillehei performed his first three cross-circulation operations and invited DeWall to observe. From his duties in the lab, DeWall already knew how to run the pump, and, soon enough, Lillehei asked him to replace the pump technician in the next open heart operations.

Dick DeWall, animal attendant, was now a full-fledged member of the clinical team.

Like the surgeons, DeWall saw firsthand cross-circulation's limitations. He knew they had to find a better method than using a human donor. One day, Lillehei talked to him about possibilities.

Dick, said Lillehei, we've got to have an oxygenator, a machine. But it has to be something simple.

DeWall got right on it.

Years later, when myth enveloped Lillehei, a story was told of how the DeWall-Lillehei machine had been born. It was no secret that Walt and his surgeons liked to stop by the bar after a long day. According to the story, the boys were downing a few one evening when Morley Cohen (or maybe it was Herb Warden or even Lillehei himself) gazed down into a freshly poured beer and said, Hey—why not try bubbles!

The real story: DeWall, in the grand tradition of American inventors, just started tinkering around.

At first, DeWall trod a familiar path: he monkeyed around with metal screens and disks, with which Dennis, Gibbon, and most others working on a heart-lung machine oxygenated blood. But screens and disks presented problems, and DeWall, guided by Lillehei's philosophy of simplicity, soon took an interest in bubbles. What were bubbles, after all, but oxygen-rich air?

Untrained in research, DeWall did not visit the library to read all the articles about failed experiments with bubble oxygenation—and there were many failed experiments. He wasn't deterred by the fact that bubbles killed tissue, nor by the conventional wisdom that bubbles introduced into blood could never be safely removed.

He merely discussed bubbles with Lillehei, who was open to anything, and then he took some tubing, some rubber corks, and some dog blood and went back to tinkering.

One of DeWall's early attempts at bubbling was a sloped tube into which blood was exposed to oxygen under pressure. This did oxygenate blood, but when the blood was decompressed, it foamed, a process not unlike what happens inside a deep-sea-diver's body when he surfaces too fast and gets the bends.

But DeWall made an intriguing discovery with his pressure system: the bubbles that formed in the collection reservoir formed on top, while the bottom blood seemed bubble-free.

That made sense. Blood with bubbles was lighter than blood without—of course it rose to the top.

Now he was onto something.

And something else, too: DeWall concluded that the size of the bubbles must matter. Many small bubbles were more difficult to remove than a few large ones, yet other researchers had used microporous glass to produce their bubbles—hundreds, even thousands of tiny, deadly bubbles.

So DeWall created larger, more visible, and less numerous bubbles—with eighteen standard intravenous needles pushed through an ordinary rubber stopper and connected to an oxygen tank.

When he was done, DeWall had built a machine with no moving parts that (excluding the T-6S pump) cost less than fifteen dollars. What a basic contraption it was, really nothing but a couple of metal stands, some large-diameter beer hose, a cork, a plastic tube, a reservoir, the needles, and two filters. A high school science-fair project was more complex.

But it worked, and this is how: Oxygen-depleted blood was fed into the plastic "mixing" tube, where comparatively large bubbles of oxygen were introduced through the eighteen needles. Carbon dioxide was released and oxygen absorbed, just as in the living lung, and the refreshed blood flowed past an anti-foaming agent (a product used in the dairy industry) and down a long, descending spiral of beer hose. On the blood's passage down, any remaining bubbles rose to the top of the stream and burst, leaving bubble-free blood to drip into a reservoir, from which the pump returned it to the patient.

DeWall couldn't see any bubbles in the blood coming out of the machine—but he knew there could be invisible ones. So he operated on ten dogs. Using the low flow rate of the azygos factor, DeWall opened the dog's chests, detoured their blood around their hearts for half an hour, closed them up, then observed them closely for days. He saw no sign of neurological damage in any. They were the same old dogs they'd always been.

DeWall told Lillehei, who insisted on seeing for himself.

Lillehei let the dogs out of their pens, walked them around, shone a flashlight in the pupils of their eyes, watched the wag of their tails, but he could find no evidence of damage, either. And talk about simple—fifteen bucks and beer hose was simple! Lillehei was so excited that he shot a movie of the first ten dogs.

DeWall tested another series of dogs, with the same results.

We're pretty close here, he told Lillehei.

It was early 1955.

———

By then, the events of the terrible fall of 1954—when so many had died in Lillehei's hands—were receding.

Starting with his second successful repair of tetralogy of Fallot, on December 3, 1954, Lillehei had sent healthy children home in fifteen of his last nineteen cases. He was so confident that he was scheduling two operations a week.

Outside of the operating room, things were heating up again.

Valentine's Day and the anniversary of Gregory Glidden's operation had prompted a torrent of new stories, and new demands for speeches and ribbon cuttings as the American Heart Association capitalized on Lillehei's successes—and the surgeon's most celebrated patient, Pamela Schmidt, the little Queen of Hearts, wore her crown. More and more hopeful patients journeyed to Minneapolis as distant doctors made more referrals, sometimes, unknown to Lillehei, collecting a finder's fee of $400. What greed! Lillehei himself charged only $400 for an open heart operation, a fee he would waive for a poor child.

Lillehei tried his best to accommodate everyone, often quietly admitting children to general pediatric wards (where technically they didn't belong) when the allotment of surgical beds, established by bureaucrats, had been exceeded, as it frequently was.

Still, the waiting list grew and grew. Children whom Lillehei couldn't reach in time were dying.

They needed the machine.

It was almost ready.

But now a formidable new competitor had joined the quest. Two hours south, at the Mayo Clinic, a schoolmate of Lillehei's was about to try his own machine on a person.

———

John W. Kirklin had been an undergraduate with Lillehei at the University of Minnesota, but it was no wonder that their paths had never crossed. After football on Saturdays, Lillehei and his buddies liked to ring in the dawn at Mitch's, but Kirklin was the furthest thing from a party boy. He was introverted and studious, a profoundly original thinker who had a lifelong passion for statistics. Kirklin believed that the vital forces could be quantified and, to a large degree, controlled.

A magna cum laude graduate of the Harvard Medical School (his one C was in surgery), Kirklin started down the road to cardiac surgery the day that he first saw Professor Gross, whose patent ductus operation had inaugurated the closed heart era. Kirklin trained at the Mayo Clinic, where his father was the chief of radiology, then at Harvard's Boston Children's Hospital, back with Gross. Like Lillehei, Kirklin was undeterred by risk. Like Lillehei, by the early 1950s he'd settled on heart surgery as his life's work.

And like Lillehei, Kirklin had considered using a second patient for open heart operations. He was at the wrong place for such radicalism, for while the Mayo Clinic had few equals in training doctors, its research philosophies were conservative—so conservative that some of Wangensteen's brasher devotees called the Mayo "charm school" (whereas they said of their own University Hospital, "You practically have to invent an operation to get on the schedule").

In any event, Kirklin thought a heart-lung machine had greater potential than a human donor. Once perfected, a machine could be more strictly controlled.

After visiting several heart surgery researchers, much as Lillehei had done, Kirklin in 1952 concluded that John Gibbon had designed the best machine. Gibbon let him into his lab, but leaving with blueprints was another matter—he'd shared his secrets before with Clarence Dennis, who'd run off and grabbed the title of first. And now another prodigy was eager to eclipse the father of the movement. But high-level discussions between the Mayo and Gibbon's Jefferson Medical College led to release of the blueprints, and a year after requesting them, Kirklin finally went into the lab. Gibbon's machine had several desirable features, but Kirklin intended to improve it.

Kirklin began his research with the assistance of Dr. Earl H. Wood, a Mayo Clinic physiologist and University of Minnesota graduate who had studied under Maurice Visscher, the university's renowned chief of physiology. Wood was a prolific innovator who one day would make a name studying the effects of space flight on the human body. When he teamed with Kirklin to develop a heart-lung machine, Wood was perhaps best known for his invention (tested on a human centrifuge) of the anti-gravity suit, which allowed military pilots to wage high-speed aerial combat without blacking out.

Even with such creative intelligence, progress on Kirklin's machine was excrutiatingly slow. Gibbon recorded his first success—and Kirklin, Wood, and their associates were still in the lab. Gibbon lost his next two patients, then gave up—and Kirklin's group was still in the lab. Lillehei brought forth the Queen of Hearts—and they were still in obscurity, still in the lab. But Kirklin and Wood were perfectionists: they would not rush into the operating room with some half-baked doodad just to prove to that Twin Cities maverick that the Mayo wasn't charm school.

Kirklin and Wood's creation (which would be known as the Mayo-Gibbon machine) was as big as a Wurlitzer organ and dreadfully complicated. Its principal components included three pumps, three reservoirs, tubing, a Lucite cone, a level-sensing device, a vacuum regulator, an anti-foam column, a pH meter, a heating element, an arterial filter, high-pressure stops, and an oxygenator consisting of fourteen twelve-by-eighteen-inch finely machined stainless-steel screens enclosed in a plastic chamber. Kirklin's machine cost tens of thousands of dollars to build and required a skilled technician to run.

By early 1955, it was finally working wondrously: nine of ten dogs had survived thirty minutes on bypass without ill effects (a stupid surgical mistake had killed the tenth).

They were ready to leave the lab.

Kirklin—precise surgeon, confirmed statistician, his mind often likened to a machine—gathered his people and said they would operate on exactly eight children.

Eight children: no more, no fewer.

The first seven could all die, and they would still do the eighth. Only through such a series, Kirklin believed, could statistically significant data be produced; only with sound data could they perfect their heart-lung machine.

———

Linda Stout's parents did not know Kirklin's cold philosophy of pioneering before they met him in early 1955. They knew only that their frail five-year-old had a ventricular septal defect, and she would die without surgery.

Even living in Bismarck, North Dakota, Edna and Howard Stout, a soil scientist, had read about Lillehei's open heart operations. The Queen of Hearts notwithstanding, the Stouts didn't like the sound of cross-circulation; it struck them as bizarre—and doubly dangerous. If the donor patient has hard luck and dies, thought Howard, who knew nothing of Geraldine Thompson, then both patients would die. The Stouts didn't bother calling Minneapolis; they got directly in touch with the Mayo Clinic, and were invited down for an examination of Linda.

A cardiologist explained the operation that Dr. Kirklin was proposing.

Well, asked Howard, have you been doing quite a few of these?

There was an awkward silence.

Actually, the cardiologist said, your daughter would be the first.

The Stouts wanted to run from his office, but they heard the cardiologist through, and then they met Kirklin, who described his machine and its success in the lab. Kirklin said he was scheduling many operations on people, which he would perform regardless of Linda's outcome. But he believed Linda would make it, and the Stouts believed him: thirty-seven years old, Kirklin was so smart and confident. Around the Mayo, they called him Boy Wonder.

On March 22, 1955, Kirklin opened Linda Stout and hooked her up to his machine. He then slit the wall of Linda's heart and peered inside. The diagnosis was correct—no nasty surprises.

Then, a very nasty surprise.

Pressure was climbing inside the machine. It shot up uncontrollably—275 millimeters of mercury, 300, 350, the gauge deep into the red. The machine seemed about to explode.

What could be wrong?

The machine seemed to be working properly. There had to be an obstruction on Linda's end.

Suddenly, the arterial line blew off.

The line—it returned blood to Linda through a major artery—slithered across the operating table. Blood spurted everywhere, and Linda's circulation stopped cold. Puzzled but calm, Kirklin investigated. Perhaps a cannula was blocked; more likely, the artery had gotten kinked. Kirklin replaced the cannula and straightened the artery, but now a new problem confronted him: air had snuck into the circuit. He siphoned it off and reconnected Linda to the machine. Blood flowed through the girl's body again.

Six minutes and twenty seconds had passed.

Six minutes and twenty seconds of no oxygen to the brain.

Five minutes would have been exceedingly dangerous.

Six, and all bets were off.

But the operating room was cool, and Linda's temperature had dropped; she'd become mildly hypothermic, and that had saved her brain. Her vital signs picked up and she seemed okay. Kirklin closed her VSD with five silk stitches, then sewed up her heart and her chest.

Linda awoke without incident and acted alert and herself. Kirklin discharged her in ten days.

Kirklin quickly completed his seven other cases. He lost four of these children, but three went home: 50 percent mortality, which Kirklin soon drove down in further operations.

Now, and for quite some time thereafter, only two people in the world were regularly performing complex open heart surgery: John Kirklin and Walt Lillehei, two Minnesotans who lived just two hours apart.

Despite all of the glittering headlines, some inside University Hospital still had qualms about Lillehei's open heart work. Certain cardiologists still refused to refer patients to the young surgeon, and Chief of Medicine Watson remained skeptical, if no longer outright antagonistic. But cross-circulation had become too successful for anyone to challenge Lillehei when he was ready to take Dick DeWall's fifteen-dollar machine into the operating room. As a courtesy, Lillehei did inform his mentor. It was May of 1955.

Dr. Wangensteen, said Lillehei, the bubble oxygenator has been very gratifying and we're planning to use it soon.

Wangensteen wished him luck. Owen was so proud of Walt. Surely the Nobel prize would be his.

176

Lillehei's cardiologist friends identified a candidate: James Frederick Robichaud, a three-year-old with sandy hair and big brown eyes from Minto, a coal-mining town in New Brunswick, Canada.

The story of little Jimmy's life was distressingly familiar. Born seemingly normal, he was still an infant when his mother noticed that his heart beat too fast—she could see it pounding, even through a shirt. The doctor assured Catherine Robichaud that babies' hearts normally beat fast. But then, when Jimmy was about five months old, another doctor, examining him for a cold, heard a murmur.

Did you know this baby has a very bad heart? the doctor said.

Catherine was shocked; except for his pulse, the pounding, and now the murmur, Jimmy seemed perfectly healthy.

When the boy was a year old, he traveled with his mother to Toronto, where William Mustard diagnosed a VSD but did not attempt correction with his monkey-lung surgery. Jimmy had two more good years, and then he developed a persistent cough, he began to sweat, his energy dissipated, and his heart swelled. The boy was dying. The Robichauds' doctor said his only chance was with Lillehei, half a continent away.

Jimmy's father, Joseph, cut hair for a living; Catherine kept the house and cared for the children. The Robichauds could not afford the airfare to Minneapolis, but the Handicapped and Crippled Children's Fund of the Junior Red Cross joined with the Minto Children's Emergency Fund, a grass-roots organization, to buy tickets for the mother and the son. Catherine packed books and toys and flew west with Jimmy in the spring of 1955. Dr. Lillehei struck Catherine as kind—even before meeting her, by letter he'd offered to waive his surgical fees if the family lacked the means. But Lillehei did not sugarcoat the risk. He explained that he wanted to use a new heart-lung machine that had worked well in the lab but never been tried on a person.

It's not something you jump into, thought Catherine. On the other hand, what if we do nothing and he drops dead next week? How could I live with myself then? After talking it over by telephone with her husband, Catherine gave her permission.

Jimmy awaited his operation at the Variety Club Heart Hospital, where the nurses marveled at his good nature, so reminiscent of Gregory Glidden's. "Very talkative and happy—affectionate," noted a nurse. Said another: "Very brave about going to sleep." Indeed, three-year-old Jimmy seemed eager to please: he made a point of demonstrating how he could brush his teeth, and he happily amused himself when his mother wasn't visiting by blowing bubbles, paging through his books, and talking on his toy telephone. When a nurse refused him chocolate milk, a favorite, because it wasn't listed on his low-salt diet, Jimmy politely said, "Ask my doctor when he comes!"

The nurse did—and Lillehei granted his wish.

On May 13, 1955, Jimmy visited Lillehei's table.

His heart lay open for seventeen and a half minutes, during which Lillehei closed the VSD with stitches and a plastic patch. The new heart-lung machine ran uneventfully.

That afternoon, Catherine visited her son in his room.

I'm thirsty, Mummy, said Jimmy.

Catherine gave him chocolate milk through a straw.

It was nearing midnight when Catherine left Variety Club Heart Hospital; Jimmy was resting easily, and Lillehei had urged the mother to get some sleep of her own. She returned to her room at the Salvation Army, which helped families of heart-crippled kids, many of whom lived so far away.

Shortly before dawn, the telephone jangled Catherine awake. The nurse told her to come quickly.

Catherine walked right into her son's room. It was mobbed. Everyone was frantic. They were trying to resuscitate her son. Dr. DeWall was there and Dr. Lillehei had been called.

Catherine fainted.

The doctors kept trying.

Cardiopulmonary resuscitation had not been invented in 1955, nor did doctors have defibrillators—electrically charged, external metal paddles—with which they might save a rapidly failing patient. All they had was manual massage: squeezing the heart with hands.

DeWall picked up a scalpel. Technically still only an animal attendant, he quickly opened the little boy's chest and began to massage his fading heart. DeWall kept the boy alive until Lillehei arrived and took over.

For another hour, Lillehei squeezed.

But Jimmy's heart would not come back.

At 5 A.M., Lillehei pronounced the boy dead.

Yet again, Lillehei had to write a grim final chapter.

But like Gregory Glidden's, the autopsy results encouraged Lillehei. Surgery had not killed Jimmy. The cause of death was endocardial fibroelastosis: a fatal thickening of the lining of the left ventricle that had not been diagnosed preoperatively, nor detected by Lillehei, who'd repaired the VSD through the usual approach, the right ventricle. Jimmy was twice cursed. He'd been born with the hole in his heart, but the endocardial fibroelastosis had probably been caused by mumps, for which there was no vaccine in the early 1950s.

The airline would not carry a dead body, so Jimmy traveled home by train. Catherine's hometown doctor insisted she not accompany him; she'd already experienced too much. She flew back to Minto to arrange Jimmy's funeral.

Satisfied that his technique and technology had proved sound—that "a most baffling disease," as he wrote in a letter to Catherine, had defeated them, not the machine—Lillehei tried again eighteen days after Jimmy.

And this time, he lost a nineteen-month-old four days after surgery.

But on his third try, the morning of July 12, Lillehei succeeded. Using the bubble oxygenator, he repaired a tetralogy of Fallot and sent the patient, twenty-month-old Jessie Weddle, home. Quickly, four more cases—and four more cured kids. By that August, Lillehei had abandoned cross-circulation.

He had his machine.

———

Not yet thirty-five, Denton Cooley already enjoyed a reputation as one of the hot shots of closed heart surgery. Cooley worked amazingly fast, and he made so few mistakes it was unbelievable. He innovated. Teaming with surgeon Michael DeBakey at Houston's Baylor University College of Medicine, Cooley had devised a way to repair aortic aneurysm, a potentially deadly weakness in the main artery off the heart. And that was only one of the significant contributions he (with DeBakey) could claim. At an age when many surgeons are just starting their practices, Cooley was a Texas legend.

But Cooley didn't have a heart-lung machine. He ached to get inside the heart, but the contraptions he'd cobbled together in his lab had accomplished little but to add to the honor roll of dogs sacrificed to science.

Cooley had heard Lillehei outline cross-circulation in February 1955 at a meeting in Houston. The Texas surgeon was floored by the Minnesotan's message and by his style. Unlike many academic surgeons, who illustrated their dull presentations with nothing but slides of X rays and graphs, Lillehei showed photos of his young patients—and they were not wearing some hospital johnny with a black bar across their eyes, like some sort of criminal, as you often saw in the medical journals. Before sending children home, Lillehei invited them by the office for a portrait. When parents asked what their child should wear, Lillehei often suggested cowboy or cowgirl outfits—little ten-gallon hats, boots, and toy six-shooters in imitation of Roy Rogers and Dale Evans. Lillehei's photos irked the old, conservative types, who held that a surgeon should be a man of science, not a showman.

But Cooley had similar flair; he himself sometimes wore ten-gallon hats and snakeskin boots. And he was amazed by Walt's results with cross-circulation, as was his partner, De-Bakey, who stood to say, "I think it is perfectly wonderful and I think they deserve a great deal of credit for their imagination, for their boldness." On that February day in Houston, Cooley resolved to visit Lillehei.

He got to Minneapolis in June. Lillehei invited him into the operating room to watch the repair of a tetralogy of Fallot with cross-circulation. That impressed Cooley, but what Lillehei showed him in the lab astounded him: the bubble oxygenator, being readied for its third try on a person.

On his way home, Cooley stopped by the Mayo Clinic to see John Kirklin's machine. Cooley had neither the money nor the staff to build or operate anything like that. But fifteen bucks for beer hose and a rubber stopper?

Working with DeBakey, Cooley first learned how to use Lillehei's new heart-lung machine on dogs; in less than a year, they used it on a person. Before 1956 ended, Cooley and De-Bakey had completed nearly one hundred open heart operations using the DeWall-Lillehei oxygenator (and a modified version designed by Cooley).

If cross-circulation brought a stream of visiting doctors to Minneapolis, the DeWall-Lillehei machine brought a flood.

Surgeons arrived from Baltimore, Philadelphia, and Los Angeles. Dennis returned from New York, and some of Kirklin's people made the two-hour trip north (as some of Lillehei's surgeons went south to the Mayo Clinic). Dwight Harken came east from Harvard, and the distinguished Lord Russell C.

Brock flew over from London. On his six-month tour of North American medical centers, Canadian George A. Trusler spent two weeks with Lillehei. "This is certainly the place to see open-heart surgery," Trusler wrote home to his boss, William Mustard, who soon traded his monkey lungs for a DeWall-Lillehei machine.

Lillehei welcomed everyone—even television cameramen, prompting one of Wangensteen's saucier residents, Norman Shumway, to observe, "Hell, I don't have to look at the operating schedule—I can turn on the TV!"

Visiting surgeons intent on learning the new procedures joined Lillehei around the operating table, while more casual observers watched from the dome, the custom-designed glassed-off gallery above Lillehei's new operating room. The dome featured an intercom, over which the onlookers could ask questions and Lillehei could answer while he worked. The hospital even provided opera glasses—and on long days, guests received box lunches so no one had to miss a minute. When the day was done, Lillehei always treated his guests to dinner or, at the very least, a round of drinks.

Lillehei was realizing his ambition of bringing advanced open heart surgery to the masses. To help visiting surgeons begin open heart programs of their own, he mimeographed and distributed copies of the bubbler's design, along with addresses and telephone numbers for the companies that made the pump, the tubing, and the anti-foaming agent. He published articles in all manner of journals. He made movies of his open heart operations and showed them at conventions.

Easy as the bubbler was to build, duplicating Lillehei's success became easier with the introduction of an improved version in 1957. It sold for less than a hundred dollars, preassembled and presterilized.

———

Before the 1950s ended, surgeons everywhere were using the DeWall-Lillehei machine (or a variation). It was by far the most widely used of all the heart-lung machines, and it hardly lacked for competition. A few hospitals purchased the Mayo-Gibbon apparatus when a manufacturer brought it to market (at a cost of several thousand dollars), and doctors in Sweden, Japan, the Soviet Union, and elsewhere built their own machines. Gross belatedly built one after finally discarding his atrial well, and Charlie Bailey got his to work.

But next to the DeWall-Lillehei bubbler, the most popular machine was one designed by another surgeon who had trained at Minnesota: Frederick S. Cross, who had assisted John Lewis on his first open heart case. After his residency, Cross moved to Cleveland, where he teamed with Earle B. Kay to build a simple, reliable, and cheap machine. But the chief of surgery at Cross's hospital feared the thing, and he forbade his surgeons from trying it there. So, on January 18, 1956, Cross and Kay loaded their contraption into a station wagon and drove it across town to another hospital, where they used it to sustain a girl's life while they opened and repaired her failing heart. More successes followed, and the machine went into wide production.

Although critics found fault with almost any machine, they saved their harshest shots for the DeWall-Lillehei bubbler. It seemed by now that anything with Walt Lillehei's name on it automatically became a target.

So you couldn't see any bubbles in the blood, the critics said—of course not! They were too small for the naked eye, but nonetheless damaging. One doubter even projected slides at a scientific meeting purporting to show microscopic bubbles in the brains of dogs he'd hooked up to a DeWall-Lillehei machine.

I don't know how the hell he gets them, Lillehei whispered to DeWall, but it's not important.

The Minnesotans' own extensive studies of dog-brain tissue had turned up nothing. And they saw no evidence of neurological damage in their rapidly growing group of human survivors.

Nor did Cooley and DeBakey find danger in the DeWall-Lillehei machine. "Using this relatively simple oxygenator," Cooley said at a major surgical meeting, "open cardiac surgery at last became clinically feasible." With the Texas pioneers behind Lillehei, the issue was settled.

———

Lillehei performed his one hundredth open heart operation—number thirty-seven with the bubbler—on December 9, 1955. But the patient died after the surgery. Death was due to heart block.

Of all the complications of advanced open heart surgery, heart block was among the most devastating. You couldn't see the nerves that make up the heart's electrical conduction system—you only knew their approximate location—and one misplaced stitch, one nick of a scalpel, and you could disrupt the heart's rhythm. The heart could slow and, sometimes, with complete block, stop for good. Often the damage was apparent before the surgeon closed the heart, but sometimes heart block set in after the patient had left the operating room seemingly cured. Repairs of VSD and tetralogy of Fallot worried Lillehei especially, for he had to place the stitches in an area rich with nerves. Lillehei calculated that 12 percent of his early VSD repairs ended in complete heart block—in death.

Unsolved, heart block would discourage the average community hospital from ever undertaking open heart surgery. The waiting lists would continue to grow, children would continue to die, and the quest would never really end.

Once again, Lillehei was shaken. Once again, he nearly recoiled from death.

But he kept his feelings bottled, as always. Even his wife, Kaye, could not open him up; she knew he had lost a child only when he came home and would not talk about his day— he wanted only a meal, a few drinks, and sleep. "When he's in trouble," Kaye later said, "you don't get to him."

Once when another child had died on his table, Lillehei

walked into the office of OR supervisor Genevieve Scholtes, a sensible woman who considered Lillehei a savior, not a murderer.

The surgeon sat down and put his face in his hands. Scholtes closed the door. Out in the hall, Lillehei's residents paced; they wanted to begin rounds.

All my patients are dying, said Lillehei.

Scholtes had never heard him sound so defeated.

Don't lose heart, said Scholtes.

Are you trying to be clever? said Lillehei.

No, said Scholtes. But you can't be discouraged. You've had a run of bad cases—but this will change.

Scholtes talked some more, and Lillehei listened. And when the woman finished, Lillehei opened the door and rejoined his residents.

Okay, fellows, he said. Let's make rounds.

———

Lillehei initially treated heart block with drugs. A combination of cardiac stimulants worked in his first three children, but only temporarily; all three died. Another drug (Isuprel) carefully administered over several days managed to save five of the next eight children with complete heart block, but 37 percent mortality was still too high.

Lillehei then turned to electricity. He knew that mankind and electricity had conducted a curious, and frequently painful, relationship over the years.

In the nineteenth century, Italian scientists had tried to revive cadavers with live current. Germans discovered that they could make the hearts of recently beheaded criminals contract by applying voltage (curious about the nature of consciousness, they'd shouted, "Can you hear me?" into the ears of suddenly bodyless heads, to no discernible response). And then there was the celebrated case of poor Catharina Serafin, a forty-two-year-old woman who had survived badly mangled chest surgery with only a translucent layer of tissue covering

185

her still-healthy heart. A German scientist experimenting on Frau Serafin found he could vary her heartbeat with current applied directly to her heart through electrodes. Poor Frau Serafin's heart yielded knowledge, but only Edgar Allan Poe could have imagined what went through her head.

The work of Paul M. Zoll seemed to have greater practical application to Lillehei. Zoll, a Harvard doctor, had in 1952 kept two patients with naturally occurring heart block alive by administering voltage to the outside of the chest.

With a standard pulse generator borrowed from the physiology department, Lillehei paced the hearts of a few of his young patients who'd developed block; like Zoll, he used metal paddles on the chest. This indeed reversed heart block, but at a horrible cost: the child had to be zapped as often as sixty times a minute, day and night—at up to seventy-five volts a zap. Even sedated, the children cried and screamed, and blisters rose and became infected.

Lillehei came up with another idea: Why not attach an electrode directly to the heart at the end of a cardiac operation? Not having to penetrate skin and bone, the current could be reduced, and the electrode could be left in place until the danger of heart block passed. With the help of Vincent L. Gott, a resident who one day would succeed Blalock as chief of cardiac surgery at Johns Hopkins, Lillehei tested his approach on dogs. At an average current of just 2.2 volts, Lillehei and Gott found, complete heart block could be reversed without pain, blistering, or any other side effects.

When a three-year-old undergoing surgery for tetralogy of Fallot developed complete heart block, on January 30, 1957, Lillehei decided the time had come to pace a person. Having completed his repair, Lillehei sewed a silver-plated copper wire to the outside of the girl's heart, brought it out through the chest incision, which he closed as usual, and plugged it into the pulse generator; an electrode attached to the skin completed the circuit. The girl lived—as did sixteen of Lillehei's next seventeen patients who developed complete heart

186

block (the one death occurred when someone accidentally dislodged the wire). Once normal rhythm had been restored, Lillehei pulled the wire out and sent the child home.

Lillehei had opened another era—but it would take him and a repairman who worked out of his garage to make the pacemaker practical.

On December 31, 1957, Lillehei successfully completed his 413th open heart surgery, by far the largest series in the world. He left University Hospital and went home to a party. It was New Year's Eve, his wedding anniversary, which he and Kaye always celebrated. They always invited all the staff surgeons, the residents, and the visiting doctors, and everyone always stayed late. It was like the good old days at Mitch's: ringing in the dawn in drunken style.

What a twelve months it had been.

The Minnesota bureau of United Press voted Lillehei Man of the Year—and newspapers seemed to publish a story a week about the pioneering surgeon. Readers in 1957 learned that Lillehei never ate lunch, that universities around the globe invited him to lecture, that his "greatest thrill" was operating on Gregory Glidden, and that he believed surgeons would soon transplant the heart and lungs. Patients from all over the world now journeyed to Lillehei. Wrote a reporter: "Attendants on the third floor of Variety Club Heart Hospital at the University of Minnesota are saying a nurse needs a speaking knowledge of at least one other language besides English these days."

As he celebrated his anniversary that New Year's Eve, Lillehei might well have marveled at the dramatic changes in the eleven years he'd been married—in more than surgery. He and Kaye now had four children, all of whom were healthy, personable, and smart. His stock investments had put him on the road to becoming a millionaire, and it showed—the Lilleheis drove Cadillacs, owned a powerboat, and belonged to a country club. Tenants of a small duplex when they married, they now owned the Binswanger House, an architecturally acclaimed residence on the banks of the Mississippi River in St. Paul.

Most of all, seven years after his lymphosarcoma, Lillehei was alive and symptom-free.

But destiny, as Lillehei and his patients could appreciate better than most, followed no script.

THE THOMPSON FAMILY, BEFORE THE OPERATION

12

One Red Rose

GERALDINE THOMPSON did not understand what she was doing that day in early 1958 when her husband, Dan, escorted her into a Minneapolis courtroom. Walking unsteadily, her left arm stiff and crooked by her side, Thompson took the witness stand. A medical malpractice lawyer began to ask her a few simple questions.

What is your name? the lawyer said.

Thompson knew it. She also knew her age, which was thirty-three.

Very good, the lawyer said. Now what about your children?

Her oldest, Thompson said, was twelve.

In fact, Leslie was eleven.

Her next child, Thompson said, was "about ten." In fact, he was nine. Asked her third child's age, Thompson said: "Oh, let me see—seven or eight." He was eight. And the Thompsons' last child? She was six, but all Thompson could come up with was "the baby."

Five minutes after she'd taken the stand, Thompson was excused. The jury needed no further proof that she was not a normal woman.

191

But she had been, witnesses testified—a vivacious, pretty young woman who had golfed, danced, and managed a family of six, including a heart-crippled young daughter. Everything had changed on October 5, 1954, in Walt Lillehei's operating room. Thompson had been about to serve as the donor for the cross-circulation repair of Leslie's VSD when an "air bubble," as the newspapers later called it, got into her bloodstream. The air did not kill her, but it irreversibly damaged her brain. That stranger on the golf course on the eve of surgery had spoken prophetically—and his words would haunt Dan for the rest of his life.

Lillehei did not cause Mrs. Thompson's injury. He didn't even see it happen: his back was to her and Herb Warden, who was readying her to be a donor. Lillehei was opening the daughter's chest when suddenly the mother's blood pressure plummeted—and Warden, who was connecting tubing to vessels in the woman's groin, saw that her blood had turned black.

Something's wrong! exclaimed Warden.

Lillehei whipped around. Seeing black blood, he immediately suspected air.

Put the right side up! Lillehei ordered.

The doctors turned the woman from her back to her left side, a maneuver that might prevent more air—whatever its source—from heading out the aorta into the body.

But Thompson's blood pressure continued to plummet. As Warden prepared to open her chest for cardiac massage, a doctor administered Adrenalin, a heart stimulant.

It worked: Thompson's heart resumed beating, and the surgeons did not have to cut her open. But Thompson was now in no shape to support someone else's life.

I don't think we can go ahead, said Lillehei.

Lillehei and Varco then closed Leslie's chest, leaving the girl's heart unrepaired.

In the weeks that followed, as it became clear that Geraldine Thompson would never be the same, Lillehei spoke confidentially to her husband, an air force major.

Dan, you have to sue us, said Lillehei. You can't say I told you that, but the money you're going to need to take care of Geraldine for the rest of her life is so great that you won't be able to handle it on your own. You have to sue.

Lillehei was right. Dan Thompson was a military man, earning only about eight thousand dollars a year.

With a lawyer, Thompson filed a $550,000 suit against Lillehei, Varco, Warden, and three anesthesiologists. The hospital offered to settle: it would pay $35,000, and Lillehei would try again with Leslie—by now, he was using the DeWall-Lillehei machine, which required no second person. Dan's lawyer wanted $50,000. Dan agonized, but finally declined to settle; he hoped for a sympathetic jury. The case went to trial two and a half years later, in early 1958. Leslie stayed sick.

Now, for the first time, Lillehei experienced bad publicity. Reporters covered the entire trial—a story every day, often on the front page, in the morning and evening papers. It was so sad, reading Dan's recollection of his wife's words when she finally came out of her coma: "Where am I? What happened to me?" It was pathetic, learning that sometimes now Geraldine believed she was in China or Bermuda, or that spiders and little men were loose in her bed, or that it was still 1954. It was a tragedy, a woman who'd once weighed 110 pounds and now weighed less than 80—an athlete and dancer who today had difficulty dressing herself, and who would wander off or nearly set herself on fire with her cigarettes if unattended.

In his turn on the stand, Lillehei, instructed by lawyers, volunteered nothing. He only answered questions: No, he hadn't seen air get into Geraldine's bloodstream, allegedly through an IV bottle that ran dry; no, monitoring IVs wasn't the surgeons' job but the anesthesiologists'. And, no, he hadn't known of any problem until his concentration on Leslie was broken by an "unusual commotion" at the table where her mother lay.

Deadlocked nine to three, the jury after almost twenty-five hours of deliberation declared a mistrial; six months later, a federal judge dismissed the case. The Thompsons did not get a penny. They didn't even get the truth.

The truth was this: An anesthesiologist overseeing the operating rooms had dropped by as Warden was tapping into Geraldine's bloodstream. The anesthesiologist saw an empty IV bottle—but mistakenly thought it was full, just not flowing. Thinking the line was plugged, the anesthesiologist gave a few good squeezes on the bulb, and air, not transparent solution, was launched toward Mrs. Thompson's brain. The anesthesiologist never publicly admitted his error. He wouldn't even have confessed to his mystified colleagues if they had not, through the process of elimination, determined what had happened.

After the trial, Lillehei continued his work at University Hospital. The Thompsons returned to Texas. Leslie finally had her open heart operation, in 1960, at a government hospital in Maryland; she survived surgery, but died of post-operative complications a few days later. Geraldine Thompson's care became too much for a man on a military salary and, reluctantly, he placed her in a mental hospital. Geraldine would cling to her own truth: that her husband had been kidnapped by agents of an enemy government, and it was her sworn duty to rescue him. Her mind would stay frozen in 1954, but her heart remained strong. At century's end, she was still alive: a reminder, more vivid than even a stone on a grave, of what pioneering open heart surgery cost.

Lillehei, of course, had abandoned cross-circulation long before the trial. By 1958, he was closing in on 350 operations with the bubble oxygenator. He was taking on ever-more complicated defects in children, and, increasingly, he was performing surgery on adults, whose heart problems were mostly acquired through bad habits, diet, and infection. He was reducing overall mortality by reducing complications.

He remained preoccupied with heart block, one of the most deadly complications.

The challenge wasn't how to pace, or electrically regulate, the surgically wounded heart: an electrode sewed to the heart wall, brought out through the chest and connected to a pulse generator, worked beautifully. Lillehei sought to replace the generator, a device the size of a tabletop radio that plugged into an ordinary electrical outlet, like a vacuum cleaner or a lamp.

Just getting a patient on this early pacemaker out of the operating room resembled some kind of quirky vaudeville act. The patient and the generator lay on the stretcher. An orderly pushed the stretcher. A second orderly followed, unreeling an extension cord as they crept down the hall. Doctors and nurses hovered while a third orderly with a second cord ran ahead to the next outlet. As the first cord reached its limit, it was unplugged and the second cord was quickly plugged in. And that was just getting to the recovery room! A generator-paced patient couldn't ride an elevator—no outlets—and what a farce stringing extension cords all the way back to Variety Club Heart Hospital, where patients recuperated, or down flights of stairs for an X ray. And you couldn't let a patient outdoors for fresh air, never mind send him home.

But those were inconveniences. Halloween in 1957 brought a dire warning. The lights went out in Minneapolis that day, Black Thursday, as local lore would remember it, and the unthinkable happened at University Hospital: the emergency power failed. The hospital engineers had installed two backup systems, just to be doubly safe—and both systems failed. Obstetricians delivered babies by flashlight and surgeons stopped operations halfway through. It was only a rumor that one of Lillehei's patients died, but what if one really had?

And what of the flip side of the coin, the pacemaker's other promise? Harvard's Zoll had already proved its use in correcting the kind of heart block that was caused by natural disease; like Zoll and other researchers, Lillehei believed that the pacemaker could be more than a temporary treatment of injury caused during surgery. He believed that many people who died from sudden, naturally occurring heart failure— from certain kinds of heart attacks—might have lived if their hearts had been preemptively regulated by a pacemaker. But unless you wished to spend the rest of your days plugged into a wall outlet praying for Black Thursday never to strike again, you were out of luck.

Lillehei needed something portable, something that ran on batteries. He knew just the man to help him.

———

Earl Bakken seemed a nice enough young fellow when he showed up at University Hospital in the early 1950s. Bakken repaired electrical equipment: centrifuges, pressure-recording devices, just about anything with vacuum tubes. He worked, in partnership with his brother-in-law, a lumberyard manager, out of a garage heated with a potbellied stove. On the side, Bakken repaired TVs.

Lillehei met Bakken during the days of cross-circulation. Frustrated by how interference from pumps and motors made it difficult to get a reliable electrocardiogram during surgery, Lillehei had gone to OR supervisor Scholtes.

Can't we get somebody up here to fix this? said Lillehei.

But the hospital electricians wouldn't come into an operating room during surgery. They said they didn't like all the blood. According to the terms of their union contract, they claimed the right to stay out.

Scholtes said she'd find someone else.

She found Bakken. Blood didn't bother him, and he soon devised a way to reduce the interference that hindered Lillehei. He was a born problem-solver, and Lillehei liked that— liked him. Outside of the operating room, the surgeon always found time to chat with the repairman, who kept a slide rule in the pocket of his flannel shirt. As for Bakken, well, he just thought Dr. Lillehei worked miracles. So it was with enthusiasm that, near the end of 1957, he went into his garage to see if he could fulfill the surgeon's desire for a portable pacemaker.

Two problems required solution: power source and size. That ruled out vacuum tubes. Vacuum tubes needed lots of juice, and they took up space.

But a new era was dawning. That miniature marvel, the transistor, was coming into its own—every issue of *Popular Electronics,* the bible of the soldering-iron handyman, seemed to sing its praises. Bakken went through back issues to April 1956, where he found instructions for building an electronic metronome in an article entitled "Five New Jobs for Two Transistors: You'll find many practical uses for these simple, low-cost circuits."

This was it—in a mass-market magazine, of all places! Powered by a 9-volt battery, this little device put out a tiny electrical pulse whose rate could be varied.

In about a month, Bakken built something for Lillehei to try in the lab. Smaller than a paperback novel, Bakken's pacemaker could be strapped to the waist or hung from the neck.

It worked wonderfully on dogs.

On April 14, 1958, Lillehei was performing open heart surgery on a child when the child developed heart block.

Bring the wire and the box, Lillehei told one of the residents.

It worked wonderfully on a person.

The portable pacemaker had arrived. And Bakken's tiny company, Medtronic, was on its way to becoming a multibillion-dollar corporation. Lillehei, one of the early investors, was among those who would profit.

Lillehei performed his one-thousandth open heart surgery on August 12, 1960. Only a few surgeons—Cooley, DeBakey, and Kirklin among them—even approached such a record.

Once dismissed by a Harvard surgeon as a medical backwater, Minnesota had triumphed in the quest.

But as he'd discovered long ago commanding a wartime mobile hospital, Lillehei could not rest. And so he aggressively forged ahead during the sixties, when the full promise of cardiac surgery finally reached the masses. Research and clinical experience led him to further refine the DeWall-Lillehei bubbler—and he began work on a new concept, the membrane oxygenator, which eventually would supplant the bubbler as the machine of choice in operating rooms everywhere. Bakken further miniaturized the pacemaker, then made it implantable; Lillehei implanted it. Lillehei developed a popular artificial heart valve, and by experimenting with cold, chemicals, electricity, and technic, he made open heart surgery still safer. Infants and octogenarians visited Lillehei's table, including many gravely ill patients whom some surgeons declined to touch. Most went home, fully cured.

By 1966, when Owen Wangensteen announced he would retire as the University of Minnesota's chief of surgery, Lillehei was no longer the daring young risk-taker.

He was forty-eight years old, a full professor for more than a decade who'd been performing advanced open heart surgery continuously for almost thirteen years, longer than anyone in the world. He was president of the American College of Cardiology, and an officer or member of some forty other societies. President Lyndon B. Johnson had honored him in a White House ceremony; he had lectured around the globe; and he had won many more awards, including (with Varco, Warden, and Cohen) the Lasker, often called "the American Nobel" since so many Nobel laureates have won it first. He'd been nominated for the Nobel prize itself more than once.

But what mattered to the worry-sick parents of some dying toddler in Peoria, of course, was only the extraordinary number of open heart surgeons Lillehei had trained as disciples and sent forth to practice: the young men (no women—Lillehei belonged to an older generation) who trained with him for months or years. Lillehei trained 138 surgeons in all, from

America, India, Israel, South Africa, Brazil, Vietnam, the Soviet Union, and thirty other nations.

Like his mentor, Lillehei had become a great teacher.

———

As he watched Walt achieve stunning success, the aging Owen Wangensteen rejoiced in his prize pupil.

His own son, Bud, had by now left for good for Spain, where he'd married a third time and fathered more children. Bud was still drinking and writing rambling philosophical treatises that were never published; he wanted nothing to do with his father except to cash the periodic checks Owen sent to help support his grandchildren. Bud's mother—Wangensteen's ex-wife, Helen—was long dead. One night in August of 1955, she left her job as a receptionist, went home, swallowed a bottle of barbiturates, and crawled into bed, where police found her body the next day. After an autopsy at Owen Wangensteen's University Hospital, the coroner ruled Helen's death a suicide.

By 1966, Wangensteen's terrible first marriage must have seemed a distant dream. Owen now had been married for more than a decade to Sally, an avid bird-watcher and bibliophile who graciously hosted the many dinner parties her husband arranged for visting doctors and prospective University of Minnesota benefactors. In the early 1960s, an allergic reaction to a vaccine had left Wangensteen partially paralyzed and unable to operate ever again, but the white-haired chief still ran his lab and chaired his treasured department of surgery—and he was eternally on the lookout for new benefactors. Accompanied by Sally, Wangensteen would impress potential gift-givers by inviting them into the dome above Lillehei's operating room to watch the celebrated open heart surgeon in action.

"Owen cared so much for all his patients—and for his men especially," Sally later recalled. "He was so proud of them."

And Owen adored Sally, expressing his affection in an unending stream of letters almost like a lovesick teen.

About to retire, Wangensteen had devoted his energies to continuing his legacy. With his wife, he began to plan a history of medicine library at the University of Minnesota, and the two of them started work on a book. But nothing would be more important for the endurance of his ideas than his successor.

Lillehei hated administration, but Wangensteen nonetheless believed that Walt was the only man for the job.

———

Many at the University of Minnesota Medical School didn't believe the rumor that began to circulate just before Christmas of 1966: Lillehei was not being named Wangensteen's successor. It couldn't be true! They couldn't be passing over Lillehei, whom the great professor loved as a son.

But it was true. The school had chosen thirty-nine-year-old John S. Najarian, vice chairman of the department of surgery at the University of California's San Francisco Medical School. Najarian specialized in kidney surgery. He was also an immunologist, but his heart surgery credentials were unimpressive.

Worse, to some, was that Najarian had played football—as a tackle, no less! And he had excelled. He had played in the 1949 Rose Bowl, and then been drafted by the Chicago Bears.

Normally an emotionally contained man, Wangensteen fired off an apoplectic letter to Robert Howard, the University of Minnesota Medical School dean. "We will be faced with a long, dry period here if Dr. N is made chairman," the retiring chief wrote. "Many will leave. You will probably be faced with the loss of Walt."

But Howard believed the medical school needed a leader from the outside, not someone beholden to the status quo. He wanted a strong administrator—a chief of surgery and department chairman who went to meetings and effectively processed paper, not someone, like Lillehei, who left it in piles until it magically went away.

And there was another concern: Lillehei's lifestyle. Everyone knew that Walt loved to party—and now the gossipmongers claimed that Lillehei, still married to Kaye, was seeing a nurse at the Variety Club Heart Hospital.

Dean Howard held firm.

"Having considered the matter from every angle as thoroughly, conscientiously—and, yes, agonizingly—as I can," Howard wrote back to Wangensteen, "I have decided to extend the offer to Najarian and to make it as attractive as I can with the unreserved hope that he will accept."

Lillehei was disappointed but not altogether surprised: his devotion to surgery and research over all else was no secret. But he still wanted the job—and he believed he could still get it. In late December, he spoke to Howard, an encounter he followed with a three-page letter. Now, Lillehei wrote, was the "obvious time to 'shift gears,' so to speak, and take on the added administrative responsibilities involved and to decrease the time spent in the operating room."

Lillehei maintained that he could become an able bureaucrat, and for evidence he pointed to his wartime command of "a 1,000-bed mobile surgical hospital with 33 officers, 35 nurses, and 750 men." Lillehei did not, of course, remind Howard of his notoriously sloppy bookkeeping—nor disclose that he was chronically late in filing his federal taxes.

Howard wanted Najarian.

Early in 1967, Najarian accepted his offer.

Courted in the past, Lillehei had always refused offers to leave Minnesota; he liked to joke that he must be slipping when search committees came calling.

But this was different: Lillehei had been slighted, and surgeons everywhere knew. Najarian tried to befriend Lillehei when he arrived in Minneapolis, but Lillehei spurned him. He didn't like the direction the new chief planned for the department, nor the mandatory weekly meetings that the new chief instituted. Lillehei looked at the young outsider and thought of how he wasn't Wangensteen, nor any of the old chief's men. Najarian's only association with the glory days was when he'd watched Lillehei perform open heart surgery—once, in 1955. Najarian watched from the dome.

That spring, Lillehei accepted an offer. He would become professor, chief of surgery, and chairman of the New York Hospital-Cornell Medical Center's Department of Surgery, in New York City. There were worse places for a surgeon who, at least in his profession, was a star.

In June, Lillehei announced he was leaving Minnesota.

Najarian wished him well, but he shed no tears. "The loss of one man does not crumble a mountain," Najarian told the *Minneapolis Evening Star.*

But it wasn't only one man who would be lost; Lillehei intended to take a dozen or so of his best heart surgeons, and he wanted his equipment—hundreds of thousands of dollars worth of equipment accumulated over more than fifteen years. Najarian could not stop people from leaving, but he maintained the equipment belonged to the university. Lillehei insisted it belonged to him, because it had been purchased with grants awarded in his name for his research.

Najarian had someone inventory everything in Lillehei's lab and then he presented the departing surgeon with the list.

These are the things that will stay, said Najarian, and these are the things that you can take.

Lillehei received the list without comment.

The Saturday before leaving for New York, in late 1967, he and several of his men rented three U-Haul trucks. Waiting until dark, they backed the trucks up to Lillehei's lab—and cleaned everything out.

They left only one thing: a red rose, in the middle of an empty floor.

CHRISTIAAN N. BARNARD

13

Moral Considerations

SHORTLY AFTER MIDNIGHT on December 3, 1967, a surgeon whom Walt Lillehei had trained unplugged the respirator that was keeping a brain-dead young woman alive. Twelve minutes later, the woman's heart arrested.

The surgeon's assistants quickly opened the woman and connected her heart to a heart-lung machine—a DeWall-Lillehei–type bubbler—which revived it. Then they began to cool the woman's body. When the heart reached 16 degrees Centigrade, the assistants cut it out and placed it in a bowl of ice-cold solution. Unlike her brain, which had been destroyed when a car struck her, the woman's heart was still perfectly healthy.

It was almost 3 A.M. at Groote Schuur Hospital in Cape Town, South Africa. The surgeon Christiaan N. Barnard now went to work on the dying fifty-five-year-old patient in the operating room next door.

Barnard connected Louis Washkansky to a second heart-lung machine and excised Washkansky's badly diseased heart, cutting along a pattern that he had learned from yet another Minnesota-trained surgeon's laboratory experiments. Barnard then sewed the brain-dead woman's heart into the man. After testing the sutures, Barnard began to rewarm Washkansky's new organ. It was now eight minutes before six o'clock in the morning.

With a single electrical shock, Barnard restarted Washkansky's new heart. He then attempted to wean the man from the heart-lung machine, and on the fourth try, he succeeded. The dead woman's heart beat strongly in its new body. At 8:30 A.M., Washkansky was wheeled out of the operating room of a doctor who was virtually unknown outside of surgery—but who would soon be a celebrity.

Word of the first successful human heart transplant astonished the world. Washkansky lived only eighteen days; pneumonia, always opportunistic in cardiac patients, killed him. But his death was anti-climactic. No operation in history had ever generated such publicity. *The New York Times* put it on page one, it led every TV and radio newscast, and Barnard flew to America to appear on *Face the Nation*. Lyndon Johnson invited him to visit, and renowned colleagues showered him with praise. Lillehei sent his old pupil a cablegram the moment he heard, as did Denton Cooley. "Congratulations on your first transplant, Chris," wrote Cooley. "I will be reporting on my first hundred soon."

It wasn't just Barnard's surgery that made headlines—the surgeon fired the imagination as well. Not even Walt Lillehei cut a figure like forty-five-year-old Chris Barnard, a handsome man with wit, charm, and an eye for the ladies. Still married to an Afrikaner he'd wed while he was only a medical student, Barnard began dating beautiful women. "After years of study and hard work," he wrote, "I needed some fun. My life was missing that spark, that extra something that makes living a joy." Barnard was soon romantically linked to Gina Lollobrigida and Sophia Loren, among other women.

His operation indeed made Barnard a star. But it was actually another of Lillehei's students who had perfected heart transplantation, during years of quiet research, and who, through still more years of dogged determination, would do more than anyone to make the procedure almost common-place.

———

Of all the young surgeons to pass through Minnesota in the 1950s, Norman E. Shumway was the most irreverent, which was saying a lot.

When a particularly arrogant surgeon once bragged that he was the only chief resident whom Owen Wangensteen had ever personally assisted on a certain type of operation, Shumway said, George, there's no chief resident who needed help more than you. Another day, one of Wangensteen's patients kept mistaking Lillehei for Wangensteen. Wangensteen was visibly annoyed that the woman didn't recognize her own doctor, but despite Wangensteen's insistence that *he* was Wangensteen, the woman could not be persuaded. Exasperated, Wangensteen turned to leave. Shumway put his arm around him. You know, said Shumway, you've *got* to stop this going around the hospital telling everyone you're Dr. Wangensteen!

And Shumway was very good with the knife. He studied for six years at the University of Minnesota, primarily with John Lewis, from whom he learned open heart surgery. He polished his craft during several weeks with Lillehei in 1956 and, soon thereafter, uninterested in spending any more time as a resident (he was married and needed money), he left for California. In 1958, he was hired by Stanford University School of Medicine, where he teamed with Richard R. Lower, a resident who also sought to advance open heart surgery. Like Lillehei, Shumway and Lower pursued parallel tracks in the operating room and in the lab.

One day, as they were struggling with a dog heart, a brainstorm struck Shumway: Why don't we just take the damn thing out and then sew it back in?

They could work on a heart more easily outside the chest. It would be like fixing a television on a workbench, not inside the cramped confines of a console.

So they began taking dog hearts out and sewing them back in.

Soon, they were taking one dog's heart out and sewing another's back in.

By the end of 1959, some of their dogs with transplanted hearts had lived as long as three weeks, an extraordinary and unprecedented achievement.

The next year, at a meeting of the American College of Surgeons, Shumway and Lower presented their "Studies on Orthotopic Homotransplantation of the Canine Heart," a paper with a long-winded title that was only two pages long. No one in the audience commented—not even a jealous whisper. It was as if something had gotten into the water and rendered the membership of the mighty college dumb. Perhaps their colleagues considered Shumway and Lower no different than the many other scientists—some visionaries and some fools— who had tried transplanting animal hearts before. Experiments had been conducted at least as early as the turn of the century, when Nobel laureate Alexis Carrel attempted to transplant the hearts and the lungs (and even both organs together) of dogs and cats; more recently, researchers at the Chicago Medical School had grafted the hearts of puppies onto the necks of their mothers. But until Shumway and Lower, no one, not even a Nobel prize winner, had succeeded in true transplantation.

Shumway and his partner returned to Stanford, where, having devised the surgical technique, they set out to solve another piece of the transplantation puzzle: the body's natural rejection of foreign tissue. Other researchers, meanwhile, were joining this latest heart quest.

In 1964, word circulated of surgeons at the University of Mississippi who, inspired by Shumway's work, had transplanted a chimpanzee heart into a sixty-eight-year-old man;

the man died an hour after being disconnected from the heart-lung machine, but that one hour suggested that the first successful human transplant was now only a matter of time. At Stanford, Shumway was ready to move from the laboratory into the operating room, but he lacked one thing: a donor. Shumway knew his chances would be best with a still-beating heart, but no Stanford neurologist would declare a brain-dead person clinically dead (even though, late at night, some of the neurologists quietly turned off the respirators of patients with flat-line brain waves). In America, the clinical definition of death was still based on a heart that no longer beat.

Christiaan Barnard did not intend to pursue a surgical residency at the University of Minnesota when he graduated from medical school in South Africa in the mid-1950s. He intended to study in London. And he planned to become a general, not a cardiac, surgeon.

A fellow South African brought him, indirectly, to Minneapolis. Surgeon Alan Thal was already training there and he had so impressed Wangensteen that the chief asked a professor friend in Cape Town to send him another South African. The professor recommended Barnard, who, enticed by Wangensteen's reputation, arrived on a frigid day in December 1955, the month that Lillehei completed his one-hundredth open heart operation. Barnard had never heard of Lillehei— nor was he much impressed when they met, at Walt and Kaye's wild New Year's Eve party.

"I thought he was a bit of a playboy—and he drank quite a bit, too," Barnard later recalled. "He didn't appear to be very serious."

Still intending to become a general surgeon, Barnard began to pursue a doctoral degree in Wangensteen's lab. One day in 1956, he passed Lillehei's lab, where Lillehei protégé Vincent Gott was experimenting with the latest heart-lung machine on dogs.

Are you free? said Gott.

Barnard said he was.

I'm alone here, Gott said. I'd like you to scrub up.

Barnard did and, that day, decided to learn the new heart surgery. "General surgery was so often destructive," Barnard recalled. "You took out the stomach, you took out the colon, you took out a lung. Heart surgery was repairing things, not destroying things."

Wangensteen didn't want to lose the promising generalist to another specialty, but Barnard eventually prevailed. He studied several months with Richard Varco, then spent a year with Lillehei—at first running the heart-lung machine, then operating. Barnard had learned, of course, that whatever Lillehei might have been on his own time, in the OR he was unchallenged king. Moreover, Lillehei was a mentor, whose support at a critical moment in Barnard's residency kept the South African from abandoning heart surgery altogether.

Barnard was opening a child's heart one day when the left atrium began to bleed. Barnard could not adequately control the bleeding; Lillehei was not yet in the room. Lillehei soon scrubbed in and quickly stopped the bleeding—by sticking his finger in the wound. But too much blood had been lost, and the child died during the surgery.

Barnard was terribly shaken. He told Lillehei that he was quitting open heart surgery—what else could he do after leaving a child dead on the table?

I can't say I'm sorry, Barnard told Lillehei. What the hell would that help? I can't bring the child back.

Did you learn from your experience today? said Lillehei.

Yes, said Barnard,

And what did you learn?

That you control bleeding with your finger.

Okay, said Lillehei. You open the patient tomorrow.

"That was the greatness of him," Barnard recalled. "The next day was a new day with new challenges and new hopes."

A fully trained open heart surgeon after three years in Minnesota, Barnard went home to South Africa, where he concentrated on his new passion—and briefly attracted attention in the scientific community when he transplanted a second head onto a dog, in 1960. Barnard remembered his roots; he returned periodically to the United States, to pick up the latest surgical techniques and to see old friends.

By early 1967, when Barnard flew back on one of his visits, Shumway's partner, Lower, had relocated to the Medical College of Virginia. Barnard stopped by to catch up on human kidney transplantation, an emerging specialty in which Lower's new East Coast partner had pioneered. But something else captivated Barnard during his visit: Lower's laboratory experiments with transplanting the hearts of dogs, a continuation of the work begun at Stanford.

Barnard read Shumway and Lower's first, two-page paper, which described the surgical technique, brilliant for its simplicity (only a portion of the patient's heart is excised, making connection of the donor organ easier).

He says he's going to go home and do it, one of Lower's assistants told Lower after Barnard had left.

How can he? Lower wondered.

That December 3, he learned.

Barnard's operation was like a starting gun heard around the world. Just three days after the South African operated on Louis Washkansky, New York surgeon Adrian Kantrowitz took a heart from a baby born without a brain and transplanted it into a heart-crippled two-week-old infant (the infant died hours later). Barnard performed his second transplant on January 2, 1968. And four days after that, with the Stanford neurologists somewhat less jittery, Shumway finally transplanted a human heart from a brain-dead woman into a fifty-four-year-old man who lived two weeks—a survival rate that Shumway would improve as he moved toward becoming the world's most experienced heart-transplant surgeon, with more than a thousand operations by his retirement a quarter of a century later. As 1968 wore on, Shumway's old teacher Walt Lillehei, now in New York, jumped into the game. And centers in Paris, London, Bombay, and elsewhere reported transplantation of human hearts.

But not to unanimous acclaim.

The issue of donor death erupted into an international fury—with doctors, theologians, politicians, philosophers, and judges all providing their impassioned points of view. For the first time in medicine, someone could live only when someone else had died—or had the someone else really been killed? Medically, surgeons agreed that harvesting a beating heart (as compared with waiting until it stopped on its own) was strongly desirable. But was unplugging a respirator-supported person murder? Was brain death really death—or sadly diminished but nonetheless still-sacred life? Might a brain-dead person miraculously wake up someday? Could surgeons hot for glory be trusted with such weighty moral decisions? And what was one to make of the events surrounding former President Dwight D. Eisenhower, who was suffering from serious cardiac disease that August? Word that he was near death prompted at least twenty healthy people to volunteer for a transplant of their own hearts into Eisenhower's failing body.

In 1968, the field that Lillehei had broken open was now at the center of a holy debate, one that made the cross-circulation brouhaha seem minor by comparison.

"We believe," wrote Shumway, "that the prerequisite revolution in moral and legal considerations surrounding the utilization of unpaired organs from patients who have suffered irreversible cerebral death is currently developing."

Indeed, it was—developing.

Shumway himself felt the fury in August of 1968, when he completed the world's thirty-third transplant—and was immediately investigated by the district attorney. The inquiry was prompted by the Santa Clara County, California, coroner's office, which maintained that Shumway had promised to notify the office "as soon as possible" when a heart donation was imminent; the surgeon did not get word to the coroner, the coroner complained, until three hours into the surgery.

But Shumway's old partner, Richard Lower, experienced far worse troubles—and resolution of them would help shift opinion to acceptance of brain death as death.

———

Lower's entrance into the human transplantation arena began on the afternoon of May 24, 1968, when Bruce O. Tucker, a fifty-four-year-old man he had never met, started drinking at a gas station in Richmond, Virginia. Tucker, a longtime worker at an egg-packing plant, fell while trying to stand up; his head struck concrete, and an ambulance brought him to Medical College of Virginia Hospital, where Lower worked.

Late that night, surgeons opened Tucker's skull to relieve the bleeding and swelling of his damaged brain. The next morning, his condition grave and apparently irreversible, Tucker was placed on a respirator. "Prognosis for recovery is nil and death imminent," a doctor wrote in Tucker's chart.

A neurologist conducted an electroencephalogram, or EEG, which measures brain waves, and concluded that Tucker's brain was dead. Doctors notified Lower's partner, surgeon David M. Hume, who in turn telephoned police to ask for help in finding Tucker's relatives; a similar call had already been made. It was now two o'clock in the afternoon of May 25.

Half an hour later, police reported that they could find no next of kin. Doctors unplugged Tucker's respirator. When Tucker did not breathe on his own, the doctors declared him dead—and, following the protocol they'd written, they called the medical examiner, who allowed the surgeons to give Tucker's heart to a stranger. Lower, Hume, and their team had

213

been anticipating the medical examiner's approval: they were already opening the recipient's chest when permission was actually given.

That evening, Lower completed his first human heart transplant. Joseph Klett Jr., forty-eight, lived seven days before succumbing to uncontrollable rejection.

Bruce Tucker's family, meanwhile, had come forward.

In fact, at least one relative had never been far away—brother William, a cobbler, ran a shoe shop only fifteen blocks from Medical College of Virginia Hospital. And even as doctors were planning to take Bruce's heart, a friend was frantically wandering the hospital corridors, trying to find where the ambulance had taken him.

William Tucker filed a $100,000 wrongful death suit against Lower, Hume, and others, and he hired as his lawyer state senator L. Douglas Wilder, who was later the governor of Virginia. For nearly four years, as the closely watched case moved toward trial, Lower's transplant program languished.

The trial opened on May 18, 1972, amid considerable tension in Richmond. By coincidence, transplant doctors happened to be meeting that week in a hotel just down the street from the courthouse. Bruce Tucker had been black. The surgeons who'd taken his heart were all white.

Although the crux of the case technically was not the legal definition of death, that question inevitably dominated the trial, which lasted a week.

In his closing argument, Wilder said that the doctors had unplugged the respirator not for Tucker's sake—not to end his suffering—but only because Tucker "was unfortunate enough to come into the hospital at a time when a heart was needed." Wilder argued that if transplant doctors got their wish and brain death became accepted as the legal definition of death, medicine risked descending into Nazism.

"The time is not too far from us," Wilder told the jury, "when men of science made lampshades of human skin."

214

A philosopher called by Lower's attorneys saw the matter quite differently. When brain function vanishes, said Joseph Fletcher, medical ethics professor at the University of Virginia, "nothing remains but the biological phenomena at best. The patient is gone, even if his body remains. . . ."

The new philosophy prevailed—the jury took just forty-seven minutes to acquit Lower and his fellow defendants.

"Having had an opportunity to discuss these questions in recent months with laymen renewed my faith," Lower said after the verdict. "Thoughtful people can come to grips with questions like these and arrive at thoughtful conclusions. . . . This will permit transplantation to continue."

Greatly relieved, Lower returned to his work. A more famous open heart surgeon, meanwhile, was not faring as well with the law. Walt Lillehei was in the worst kind of trouble.

WALT LILLEHEI LECTURING, MID-1970S

14

New York Medicine

LILLEHEI HAD JUST swept into New York City in a Jaguar
XKE, three trucks full of the finest cardiac equipment
close behind, when Christiaan Barnard transplanted the first
human heart. Lillehei was beginning his tenure as chief of
surgery at the New York Hospital-Cornell Medical Center.

Surgically, Lillehei had arrived with style, too. His name
was barely on his office door when an allegedly hopeless case
came to his attention. Born with a horribly deformed heart, an
infant named Beth McDonough had already visited New
York-Cornell cardiac surgeons, who declared that the baby's
heart was beyond repair and she would soon die; the best they
could do, they said, was make Beth comfortable in what little
time she had left.

The new chief offered to take the case—at Minnesota, he'd
heard similar gloomy prognoses, then defied them by performing surgery that others deemed too risky. Proving colleagues wrong had never won Lillehei many friends, but
Lillehei viewed life through a different lens. Beth's parents, a
New Jersey couple, gave him their permission to operate on
their daughter that December of 1967.

Not even the world-famous surgeon could cure little Beth McDonough, but Lillehei alleviated her suffering. She lived until the age of four, when, her heart condition worsening, she died of complications from the flu. "Because Dr. Lillehei was willing to forge ahead," Beth's mother, Mary L. McDonough, later said, "he gave our family and friends the gift of enjoying a very special little girl."

As the new year began, Lillehei opened a laboratory and began to schedule a full load of open heart operations.

Cardiac surgery in the late 1960s was expanding in directions that would have surprised all but the most far-sighted pioneers—and not just transplantation, which generated the most publicity. With repair of such once-dreaded defects as tetralogy of Fallot virtually routine, surgeons now defeated many of the most exotic killers. They could even rejuvenate a heart laid low by smoking or a bad diet. Such habits can narrow the small coronary vessels that deliver blood to heart muscle; over time, the clogging starves and kills muscle, precipitating a heart attack. Once powerless, surgeons in the late 1960s devised a way to replace diseased coronary vessels. They invented lifesaving coronary bypass surgery—"bypass," as ordinary folks would call it.

Lillehei remained at the forefront of these endeavors, developing in his Cornell lab new valves, pacemakers, and surgical techniques. Spurred by his ex-pupils Shumway and Barnard—and by his early supporter Cooley, who transplanted human hearts and was developing an artificial heart, as was DeBakey—Lillehei established an organ-transplant program at the New York Hospital-Cornell Medical Center.

Lillehei's first human heart transplant (the world's twentieth) ended in failure on June 1, 1968—but in 1969, Lillehei enjoyed a series of spectacular transplant successes.

The first was in February, when he led a group of surgeons who harvested six parts of a brain-dead man and transplanted them into six separate patients—operations that dazzled the nation. "Surgeons said the operations marked the first time that as many organs—heart, liver, both kidneys, both corneas—were removed from a single donor for transplantation into six recipients," reported the Associated Press. "The operations also marked the first time a heart was removed in one hospital and transferred, via a block-long tunnel, to another hospital for implantation."

The headlines continued all year, culminating in the wondrous events of Christmas morning—when Lillehei, in what the tabloids called the "Christmas present operation," transplanted a heart and both lungs into a construction worker who was dying of emphysema. Only the second surgeon ever to attempt such a feat (Cooley was first), Lillehei brought his patient out of surgery in less than three and a half hours.

"Is the operation over?" the patient scribbled on a pad when he woke up that Christmas afternoon.

"Yes," Lillehei wrote back.

"God bless you all," wrote the patient, a middle-age man who lived another eight days, when rejection complicated by Lillehei's old nemesis, pneumonia, claimed him.

And who was unmoved reading of Dr. Lillehei's offer to fix—for free—the heart defect inside a dying German shepherd puppy owned by eight-year-old twins from Brooklyn? Lillihei learned of the sick puppy reading the boys' letter to the editor, in which the twins criticized their veterinarian, who said that the dog should be put to sleep.

"Dogs have been the backbone of open-heart surgery," Lillehei explained to reporters. And so he offered his services "on behalf of the many dogs who have served to benefit mankind in helping develop these techniques."

The puppy died before Lillehei could operate—but not before newspapers everywhere published a picture of the grateful boys hugging their lovable heart-crippled pet.

Away from the publicity, a deeper drama unfolded: Walt Lillehei seemed to be indulging sybaritic desire, as if he no longer could resist life's temptations.

Even before Walt left Minnesota, his personal affairs had become tangled. Fire nearly destroyed the Lilleheis' home in the summer of 1967, and then a boating accident nearly ruined Kaye Lillehei's beautiful face. Walt was piloting his powerboat one evening on the Mississippi River, and Kaye was sitting on the stern. The Lilleheis had been out to dinner, they'd had a few drinks, and now Walt was driving the way he loved, in a boat or in a car—too fast. He did not see the sandbar in the dark. The boat slammed into it, and Kaye was catapulted. Her face hit the dashboard, sending part of a mirror into her skull and a key through her nose; the entire right side of her face was crushed. Doctors were able to repair all of the damage, but Kaye spent ten days in her husband's University Hospital, during which time she thought to herself, Things come in threes. And they did: a few weeks later, the Lilleheis' first child and only daughter, Kim, ran off and got married. She was only nineteen.

After considering relocating with her husband to New York, Kaye decided to stay in St. Paul so that the three younger children could remain in the local schools. Walt took an apartment on New York's fashionable Upper East Side and hired an interior decorator to outfit its four bedrooms, recreation room, and split-level living room, which had a magnificent view of Manhattan. The decorator hung many of Lillehei's citations and photographs of the great surgeon meeting a president and a pope—alongside a large painting of a nude.

"Walt, don't you think you'd be better off with a landscape?" said Owen Wangensteen on a visit.

But the nude reflected Lillehei's New York lifestyle. Fresh from an affair with a Variety Club Heart Hospital nurse, Cornell's new chief of surgery began seeing other women. He threw lavish parties, inviting residents, nurses, operating-room technicians, and just about anyone else who wanted to cut loose. He hung out at bars, including the Recovery Room, which was close by the New York Hospital; and La Chansonette, where he enjoyed Marlene Dietrich recordings, which reminded him of Europe during the war. One night in a scuffle with a drunk, the distinguished professor and chief of surgery suffered a black eye.

"He just sort of went beserk," Kaye would later recall of the New York period, when she mostly stayed in St. Paul.

———

Unbeknownst to Lillehei, some in high places at Cornell and the New York Hospital had been leery of appointing him. They didn't trust Wangensteen's glowing recommendation, for they knew that when it came to his men, Walt especially, the old chief was anything but objective. They questioned Lillehei's administrative abilities. And they were concerned by reports of Walt's vices—the enduring spirit of Mitch's, the old bottle club with the fabulous Dixieland bands. But they'd been willing to give him a chance because if anyone could make Cornell an open heart power, surely it was Lillehei.

His arrival had done nothing to reassure his New York detractors. All that state-of-the-art equipment, according to the University of Minnesota, was stolen. Some of the New York surgeons resented Lillehei's importing a dozen or so doctors from Minneapolis: his "entourage," the New Yorkers called them, to which Lillehei would jokingly say, Well, you can either look at us as a metastasis or a transplant! And the penny-pinchers objected to Lillehei redecorating his office, including covering the walls of his private john in expensive fabric. "Most of us figured white paint was pretty good," said one Cornell doctor.

Nor did the New York Hospital-Cornell community warm to the little stunt Walt pulled at his first surgical grand rounds, a lecture to which all are invited. Taking his cue from Cooley, who had been known to titillate a scientific audience with a risqué slide, Lillehei during his first lecture slipped in a color photo of a bare-breasted *Playboy* bunny. Such fine outcomes were possible with excellent open heart surgery! Lillehei jokingly asserted.

Except for Walt's Minnesota boys, no one chuckled. The women and men in Lillehei's audience were horrified.

Even in his open heart surgery, many at Cornell considered Lillehei unconventional, if not dangerous. Lillehei often let assistants open the patient's chest, not himself arriving until the heart was exposed—a custom at some hospitals, but a scandal then at Cornell. He played a radio while he operated, also scandalous. And, as he'd proved early on with baby Beth McDonough, he disregarded the advice of other doctors when he believed they were being unnecessarily timid; in turn, some of those other doctors believed that Lillehei was knife-happy. Lillehei and his Minnesota-trained surgeons, said one Cornell doctor, "were just very aggressive. They would jump to operative conclusions before the studies that customarily had been done at Cornell." This was the same criticism that Cecil Watson used to level at Wangensteen and his protégé.

There was more. Lillehei immediately began to reform Cornell's residency program, which, like the Mayo Clinic during surgeon John Kirklin's early years, was heavy on clinical experience and light on the lab. Lillehei sought to bring that balance more in line with the Minnesota model, but hard-line Cornellians resisted. Who said surgeons were also supposed to be scientists, in the first place?

Walt's public persona did not always thrill the New Yorkers, either. Flattering headlines about double transplants were well and good, but Dr. Lillehei was more than the dutiful, albeit heroic surgeon; Walt spoke his mind, as always. In describing the difficulties of rushing donated organs from Memorial Sloan-Kettering Center to New York-Cornell, for example, Lillehei told reporters that the tunnels connecting the two institutions were "very formidable, like the sewers of

Paris"—a comparison that riled his detractors, who did not believe that *sewer* was an appropriate word to be used in connection with their venerable institution.

Worse, Lillehei dipped his toe in treacherous political waters. Opened in 1769, the New York Hospital was New York City's oldest, and its trustees included a Whitney, a Rockefeller, and a Pratt—rich Republicans. Lillehei himself was a Republican, but as a favor to an old friend, Hubert Humphrey, the surgeon agreed to chair N.Y. Physicians for Humphrey-Muskie, a political action group that supported the Democrats' 1968 presidential ticket. Writing on the New York Hospital-Cornell Medical Center letterhead, Lillehei sent a letter to every doctor in New York State urging money and votes for the two liberal Democrats. In his letter, Lillehei praised Humphrey's record on civil rights and disarmament, and maintained that Humphrey and Muskie "are the most likely to produce a peaceful settlement in Vietnam."

Wrote Lillehei: "Time is short. So please do everything you can *now.*"

Perhaps Lillehei could have survived if he'd been a clever bureaucrat. He was anything but.

Lillehei still hated paperwork. He still hated meetings, and he rarely attended sessions of the medical school's executive council—a shocking slight, in the opinion of other departmental chairmen and the dean. When he himself had to call a meeting, Lillehei scheduled it for late afternoon, when attendees, tired by a long day, would want to adjourn quickly. Lillehei did not spend the time with interns and medical school students that the dean and hospital director desired, and he traveled more often than they liked—twenty-two trips as visiting professor in Mexico City, Copenhagen, Miami, Madrid, and elsewhere in less than two years.

In short, Walt was what he'd always been: his own man.

By early 1970, only a year after the fantastic "Christmas present operation," Lillehei's latest detractors wanted this man out.

Kept informed by friends in New York, Wangensteen, retired back in Minneapolis now, was increasingly alarmed. In March, he wrote Walt a three-page letter filled with advice and encouragement. "You are as competent a person as I have ever known," the old chief wrote to his old pupil, "and with daily attention to your multiple responsibilities, you can, I am very certain, surmount any and all problems."

Wangensteen implored his New York-Cornell friends to save Lillehei. "Walt, of course, is a non-conformist," he wrote to one, "but a person of extraordinary ability as well as achievement."

But it was too late.

One day in April 1970, the Cornell dean and the New York Hospital director summoned Lillehei.

As of July 1, they said, they were removing him as department chairman and chief of surgery. He could stay at the hospital, continue to perform surgery, and keep his professorship, but his tenure as top dog was over. There would be no second chance, no appeal. Lillehei was to vacate his office immediately and make way for an interim chief.

He did vacate his office, leaving one red rose when he moved downstairs.

———

Lillehei was fifty-one. Sixteen years had passed since Gregory Glidden, the first cross-circulation patient, twenty since Wangensteen had operated on Lillehei to save him from cancer. Losing the chairmanship was a bitter defeat, but from the administrative angle, Lillehei was relieved. Walt still had his operating room, his lab, his devotees, his health. Twenty more productive years, easy. Life would go on.

It was about to take another incredible turn.

WALT LILLEHEI, 1997

15

Bread Upon the Waters

THE HEADLINES on April 13, 1972, stunned people who knew Dr. C. Walton Lillehei only as a world-renowned surgeon. A grand jury in St. Paul, Minnesota, had indicted Lillehei on charges of evading $125,100 in U.S. income taxes. He faced up to twenty-five years in prison—and the loss of his license to practice medicine.

Lillehei had tried to stave this off, but the Internal Revenue Service had no interest in accommodating the doctor whom some were beginning to call the Father of Open Heart Surgery. The agency wanted to make an example; it wanted coast-to-coast headlines on the eve of Tax Day, and it got them.

Had you visited Lillehei's office back in the hectic 1950s, you might well have predicted tax trouble.

The University of Minnesota frowned on doctors using the medical school's secretaries for billing, so Lillehei did his own—or didn't do his own. Bills to patients and insurers were late going out, and when the checks finally came in, Lillehei tossed them in a drawer. When he needed money he'd fish around and pull out a check that was about the amount he needed and take it to the bank—if the check hadn't expired.

When he got around to bookkeeping, Lillehei used index cards filed in shoe boxes.

And what were taxes, really, but even more paperwork? Walt, legendary procrastinator, was always late filing his returns—by a few months at first, but soon he'd gotten a year or more behind. Initially, the IRS failed to notice. Then, on April 27, 1966, Lillehei filed returns: not for the preceding year, however, but for 1963 and 1964. His income had increased, and this time, the IRS looked into his delays.

An inspector concluded that while Lillehei was a procrastinator, he wasn't a crook.

"This guy is clean," said the inspector.

But now the IRS was watching closely. When almost two years passed without further word from him, the agency sent a letter, in December of 1967, the month Christiaan Barnard transplanted the first human heart.

"Dear Dr. Lillehei," said the agency. "The files of the district director of Internal Revenue, St. Paul, show no record of federal income returns in your name for the years 1965 and 1966. We are assigning this matter to a special agent."

Lillehei responded immediately, promising returns and payment by January 15. But once again, he procrastinated. It was now early in his tenure at the New York Hospital-Cornell Medical Center; his time was scarce, and he didn't file his 1965 return until November of 1968—two and a half years late. By then, Lillehei should also have filed his returns for 1966 and 1967, but he hadn't. Nor did he offer an excuse; he simply ignored the government, again.

The special agent was Ray W. Jackson, a career bureaucrat whose education consisted of high school and correspondence courses. He called Lillehei in late December 1968, but the surgeon was not available. He was busy transplanting a heart and two lungs—busy with the Christmas present operation that made headlines around the world.

Jackson persisted, and Lillehei finally began to get the picture. He hired a tax lawyer and, in May of 1969, filed returns for all three delinquent years, along with payment of taxes, interest, and penalties. But Jackson was only getting started. The special agent burrowed into shoe-boxed index cards, microfilmed bank records, travel and expense reports, all manner of receipts. He tracked down former patients, professors who'd paid Lillehei to lecture, people who'd tended bar at Lillehei's parties.

Special agent Jackson found more than evidence of tax fraud. He discovered that Lillehei had been leading something of a secret life—a life that included mistresses and even a Las Vegas call girl.

The day in 1972 that his indictment hit the news, Lillehei sent a letter to Wangensteen, who was still saddened by Walt's fall from grace at New York-Cornell.

"I know you must be disturbed by the recent tax publicity," wrote Lillehei, "but all I can tell you at this time in this communication is that it is not at all what it appears on the surface. I am quite confident that I am going to be acquitted."

Wangensteen wasn't so sure; he knew that juries could be unpredictable, and he urged Walt to settle out of court. "Washing linen in public is never a very rewarding experience, as many who have tried it have learned to their sorrow," said Wangensteen. But the IRS wasn't dealing. The agency had already given Lillehei this choice: plead guilty and put yourself at the mercy of a judge, or go to jury trial.

The prosecution began its case on January 15, 1973, with a thunderous attack on the open heart surgeon. It seemed to Lillehei that the prosecution intended to do more than prove that he'd underpaid his taxes. It seemed that the government of Richard Nixon wanted to destroy him—in retaliation, Lillehei suspected (but never did prove), for his support of Nixon's opponent in the 1968 presidential election.

In his opening statement, U.S. District Attorney Robert Renner accused Lillehei of double billing. He maintained that income from at least 318 patients, several visiting lecture-ships, and savings accounts had never been reported. He said that Lillehei had claimed the cost of his parents' fiftieth wedding anniversary party as a business expense, and gifts to three girlfriends and the Las Vegas call girl as tax-deductible "typing" expenses. As the trial progressed, Renner planned to have these four women testify about the definition of typing.

But first, he began to call former patients—people who'd paid Lillehei fees that the government claimed he had not reported. Dutifully, these upstanding citizens paraded to the stand, their stories punishing the man who'd literally held their lives in his hands. Lillehei sat quietly, sometimes looking puzzled or pained. The days of the Queen of Hearts seemed a very long time ago.

Then, on the second day, unexpected sympathy.

Carl Schuler, of Waterloo, Iowa, was testifying. Renner coaxed from him the story of how, for three hundred dollars, Lillehei had cured his wife, Shirley Schuler, who was also in the courtroom.

Renner finished and handed the witness over to the defense.

"Mr. Schuler," said one of Lillehei's lawyers, "what is your occupation?"

"Brick mason," said Schuler.

"How did Mrs. Schuler happen to go to Dr. Lillehei for services?"

"Our family physician wanted her to have the best and he is the one that contacted him."

"Were you subpoenaed to be here today?" said the lawyer.

Schuler said yes.

"Was your wife also subpoenaed?"

"Yes."

"Did she respond to the subpoena?"

"She told them she would just as soon not," said Schuler. "It was very hard for her."

As Shirley Schuler left the courtroom, the reporters asked

her about Dr. Lillehei. "If it wasn't for him," she said, "I wouldn't be here today."

The trial entered a second week, then a third.

Through a string of witnesses, Renner exposed the most intimate details of Lillehei's life. Under "drugs and pharmaceutical charges," Lillehei had deducted veterinary bills for the family cats, Tinkerbell and Peter Pan. He had written off piano lessons for his children; dance lessons for him and his wife, Kaye; and tuition for Kaye's university courses, including one on existentialism.

But what really got Minnesotans talking was the doctor's other women, each of whom Judge Philip Neville permitted on the stand over Lillehei's lawyers' objections. One was the Variety Club Heart Hospital nurse: she testified that she'd had an "intimate" relationship with Lillehei, during which he'd given her five hundred dollars, an amount he claimed on his returns under "additional professional, secretarial and maid help." The most devastating of the four paramours was the Las Vegas call girl, who said the hundred-dollar check Lillehei had given her was not for typing. The call girl left quite an impression. Judge Neville interrupted her testimony early on to ask her to please stop chewing her gum.

The question that no one could seem to answer was: Assuming that all of these allegations were true, how could such a brilliant man have been so foolish? The best answer anyone could provide was that Lillehei, in a rush to reconstruct so many years of such atrocious record keeping, had monumentally blundered.

Lillehei's billing system was the core of the government's case. Even if the jury accepted Lillehei's unorthodox accounting, argued the U.S. attorney, why was so much patient income unrecorded—and thus unreported? According to Lillehei's system, every payment from every patient was supposedly entered on one of those shoe-boxed index cards. But there were no cards for at least 318 patients.

The trial was under way when Lillehei, digging around in his untidy cellar, found more boxes. The cards therein were a mildewy mess—firefighters extinguishing the 1967 fire had caused water damage—but there they were, the missing records for the 318 patients. This couldn't be fraud, the defense maintained; this was only classic paper chaos.

Jerry Simon, Lillehei's lead lawyer, managed to get the cards entered into evidence without having to put his client on the stand. Simon now thought acquittal was likely. If Lillehei was guilty of anything, Simon believed the jurors would decide it was bad judgment, carelessness, and of course procrastination—but not fraud.

The trial had entered its fourth week when the prosecution dropped a bomb.

Unbeknownst to Lillehei's lawyers, prosecutor Renner had sent some of the newfound cards to a documents examiner; with an infrared viewer, the examiner had discovered that many had been altered. Someone using different ink had changed numbers, making it appear as if several accounts still had outstanding balances; under his unique accounting system, Lillehei wouldn't have considered payments to such accounts as reportable income (not until they'd been paid in full). The alterations seemed damning evidence of criminal intent.

Simon was floored. He offered no explanation.

The trial ended its fourth week. Renner rested the government's case, and Simon began the defense.

Treading carefully through the issue of altered cards, which he knew the jury found deeply disturbing, Simon played for sympathy. He sought to create in jurors' minds the image of Walt as a great humanitarian whose contributions had come at great personal cost; if he'd cheated anything, it had been death. The legendary Professor Wangensteen, seventy-four, could speak most eloquently to that point, so Simon called him.

The retired chief of surgery testified about diagnosing Lillehei's lymphosarcoma—about how eminent pathologists had predicted that the young surgeon would probably not survive five years and how Lillehei had thus first cheated death, his own. Wangensteen then went on to discuss something even more miraculous: the untold thousands of children and adults whom Lillehei had saved with his open heart advances. But miracles do not come free, Wangensteen testified, and in Lillehei's instance one cost was an astounding inattention to paperwork. But there were only so many hours in the day, said Wangensteen, and Lillehei had spent most of his on surgery, research, and teaching; in the balance, wasn't that what counted?

Lillehei's former partner Richard Varco also testified for the defense, along with Maria Ramsay, a wealthy benefactor who'd brought some 140 impoverished children to Lillehei, who fixed their hearts without fee. An aerospace engineer testified that Lillehei was the only surgeon who had the "guts" to perform a coronary-bypass operation on him when he was dying; thus snatched from death, the engineer had survived until Norman Shumway gave him a healthy new heart in an operation at Stanford.

"The doctor saved my life," said the engineer. "I want to save his."

There was a final twist, mere hours before the case went to the jury.

Lillehei hadn't underpaid his taxes, Simon argued—the government owed *him,* $53,000, to be exact. That was the net result of charitable deductions that Lillehei could have claimed, but did not, for royalties on a commercially successful heart valve he'd codeveloped in the 1960s. Lillehei had not profited from the valve; he had directed the proceeds to the University of Minnesota.

On February 14, 1973—Valentine's Day—the case of the United States of America versus C. Walton Lillehei went to the jury. The trial had consumed twenty-one days and cost hundreds of thousands of dollars. One hundred and sixty-four witnesses had been called by the prosecution, sixteen by the defense. More than six thousand exhibits had been entered into evidence.

"I have the definite impression that the jury is with you," said Wangensteen in a note to Lillehei. Lillehei himself was optimistic that the bad times, which had really begun long before his indictment, were about to end. Lillehei retreated to his St. Paul home to await the verdict.

His wife accompanied him.

Kaye had left town for the trial: she did not need to see Walt's embarrassing transgressions splashed in lurid detail across the hometown papers. Nor would she have learned much, sitting in court. Quite some time before 1973—during Walt's tenure at Cornell, in fact—Kaye had discovered his secret life.

Early in their marriage, Kaye gave up a future in nursing for a young family and a husband who worked six and often seven days a week. "R.N. earns MRS. Degree," declared a *Minneapolis Star* headline on an article about the celebrated surgeon's wife in 1955, when the Lilleheis' four children were all younger than seven. "What he's doing is as important as anything can be," Kaye told the writer of the story.

But the long hours hurt; even on Saturday nights, on their way to the Parker House for an evening out, Walt would insist on stopping by University Hospital or Millard Hall to check on a patient or the progress of an experiment. Walt missed his daughter's high school graduation, and he was no Little League dad to his three sons. Kaye raised the children; as they grew, she golfed, bowled, skied, and volunteered for the Red Cross, pursuits that did not involve Walt. "I created my own life," said Kaye years later. "Like they always say: the doctor's wife will either drink herself to death, or she'll commit suicide—unless she does something else."

When she discovered what her husband was up to in New York, Kaye considered divorcing him. But she did not. Lillehei lost his job as chief of surgery, and his tax troubles soon began; the party ended, and Walt begged for one more chance.

"If you can forgive me," he said in a letter to Kaye, "I'll do anything to make up for it."

———

Two days after the jury got the case, the Lilleheis' phone rang. The eight men and four women jurors had decided.

Accompanied by their daughter and Walt's aging parents, the Lilleheis went into the courthouse. Several of Walt's friends were there, including Wangensteen and Varco.

Everyone took a seat, and count one was read.

How say you? asked Judge Neville.

Guilty, said the foreman.

Count two?

Guilty.

Guilty on all five counts.

Lillehei faced twenty-five years of hard time and a $50,000 fine. The Father of Open Heart Surgery was a felon.

Walt embraced his wife and daughter. On his way out of the courtroom, he said nothing to the reporters.

Kaye said only, "I don't believe it."

———

On the morning of May 4, Lillehei, his family, and some friends returned to court. Lillehei was to be sentenced.

Jerry Simon pleaded for leniency. He spoke of Lillehei's service to his country in war—the Bronze Star he'd earned for valor. He spoke of Lillehei's generosity to patients who could not afford his services, and of his priceless advances in open heart surgery. "Humanity has benefitted immensely from the work of this man, work which was not done without great sacrifice on his part," said Simon. "Society is indebted for what he has done, and perhaps this is the time for society to recognize and perhaps in some measure repay him for the contributions he has made."

"Dr. Lillehei," said the judge, "do you wish to make any statement to the court?"

Lillehei did.

"First, of course, I stand before you labeled as a criminal," he said. "I must say I don't feel like one. I hope I don't look like one. I don't believe I am one. My conscience is quite clear on that score. I certainly have been guilty of poor judgment, a good deal of carelessness, some good honest mistakes. But those are hardly criminal acts. And I find it hard to accept the fact that I stand before you in this position."

Lillehei spoke briefly of his extensive charity, and of his many gifts to medicine—all the machines, techniques, and instruments he developed or codeveloped but which had never enriched him, since he never sought patents in his name.

"I find it difficult, if not impossible, to comprehend the charge and conviction of me being a money-hungry, conniving individual," said Lillehei.

The surgeon closed with an allusion to Ecclesiastes.

"Your Honor," he said, "I think I have cast a little bread upon the waters, and I can only hope—perhaps pray is the better word—that the time has come. Thank you."

Philip Neville was no hanging judge. He had a strong record on civil rights, and lawyers considered him a fair jurist who took his power as a sacred charge. He had also, unknown to anyone but his family, recently been diagnosed with leukemia. At the age of sixty-three, he was dying.

Few if any cases had so confounded Neville as Lillehei's. After the guilty verdict, letters poured into his chambers; they ran about fifty-fifty, with half asserting that Lillehei's contributions should spare him prison, and the other half maintaining that no man, Nobel contender or not, was above the law. Neville was similarly torn; for nearly three months, he had agonized over what to do.

In the end, Neville decided to fine Lillehei the maximum, $50,000, and order him to serve six months of community service. Neville could not bring himself to imprison Lillehei.

"I can't help but recognize that you have this great talent that should be of use to society," the judge said. "And sitting in jail in a cell, if it doesn't destroy that ability—if you are there for several years, at least it impedes it and renders it [useless] for that period of time."

After getting away with Kaye for a few days to Hawaii, Lillehei returned to the New York Hospital-Cornell Medical Center, where he was allowed to keep his operating privileges through the end of the year. After that, the director and the dean wanted him out altogether. They would honor the financial terms of his contract, which ran through the end of 1974, but that final year, they decreed, would be a paid leave of absence. Come December 31, 1973, they wanted Lillehei off the premises.

They needn't have bothered.

Lillehei was developing cataracts—the cruelest possible side effect of his radiation treatment for cancer was manifesting itself almost a quarter of a century later. There is no good time for a surgeon to suffer such a curse, but for Lillehei, the timing couldn't have been worse.

On December 30, 1973, Lillehei performed open heart surgery. Then he left the New York Hospital without telling anyone of his suspicion that he'd done his last operation.

Lillehei was scheduled for cataract surgery in 1974, but he doubted any ophthalmologist could restore the sharp eyesight he needed for his work. He was right. Denton Cooley would still be operating well into his seventies, and Michael De-

Bakey would still be going at ninety, but C. Walton Lillehei, fifty-five years old, never scrubbed again.

———

Lillehei's tribulations did not end with his trial. Heartbroken, some said, by their son's humiliation, Lillehei's mother and father both died before 1973 ended. The American College of Surgeons suspended Lillehei indefinitely, and the state of Minnesota revoked his license to practice medicine there until completion of his community service.

But Lillehei had trouble even finding a hospital that would let him fulfill his sentence: surgeons, including some Lillehei had trained—chiefs and department chairmen themselves now—shied from association with public scandal. So what if, for some, this was hypocrisy? Lillehei's private affairs, not theirs, had been all over the front pages.

Lillehei could not bring himself to beg for a job, but Wangensteen could.

Wangensteen wrote letter after letter, pleading, calling in old favors, praising Walt's brilliance. "As I said at the trial, he is truly one of the Surgical Immortals," Wangensteen wrote to one prospective employer. "To the harsh impeachment of C. Walton Lillehei I would remind his critics of that ageless truth that to err is human, to forgive divine."

But not even Wangensteen could change perception.

"Gossip is so rampant that Walt would have a very difficult time of it here," wrote the chief of heart surgery at one prominent center. "As a matter of actual fact, I know I could never get him appointed. The only thing worse would be if I did. Then there would be nothing but gossip and innuendo."

A veteran's hospital in Brooklyn finally agreed to let Lillehei serve his community service there—but not without a fight. Wangensteen prevailed only through Senator Hubert Humphrey, who intervened on behalf of his surgeon friend.

———

On October 6, 1975, nearly five hundred heart surgeons and cardiologists gathered at the Henry Ford Hospital in Detroit. The invitation-only Second International Symposium on Cardiac Surgery was the most prestigious such affair in twenty years. Every living leader in heart surgery attended or sent an original paper: Cooley, DeBakey, Shumway, Lower, Dennis, Bigelow, Mustard, Harken, even the old renegade Charlie Bailey.

Yet two big names were absent.

Chris Barnard was one. The invitation committee believed that the South African surgeon had plagiarized the work of Shumway and Lower, even though Barnard had always acknowledged the debt he owed them.

Lillehei was the other.

Walt had attended the first Henry Ford Symposium, in 1955, when he electrified the audience with a masterly presentation of surgical correction of VSD, tetralogy of Fallot, and atrioventricular canal. His name had come up during selection for the second symposium, of course. But led by a longtime critic of Lillehei—a surgeon whose own meager contributions did not match his ego—the invitation committee had voted Lillehei down.

Another rebuke by his peers; by then, they were common. Lillehei bore his new status as pariah with humor and grace. He was sensible, not vindictive; sanguine, and only rarely angry. Lillehei had been places most never go, seen things most never do.

He thought, *Isn't this life, too? Isn't the guy on top of the mountain always a target?*

Since before his breakthrough with Gregory Glidden, even, hadn't the critics sniped? Not those, like the Mayo Clinic's John Kirklin, who disagreed with this or that surgical approach because they believed they had something better. Not those critics, but the ones who resented the publicity, the parties, the Jaguar XKE, the architecturally acclaimed house—and mostly, the shimmering genius behind the piercing blue eyes.

Lillehei had cast bread upon the waters, all right. In 1975, nothing was coming back.

———

So Lillehei went home to Minnesota. He watched two sons become surgeons, and the third a businessman, and his daughter a good mother and wife. He did not become a recluse—he still dined with Kaye at the Parker House, still socialized with friends at the Pool and Yacht Club. He paid his fine, but he was hardly left impoverished. He was a millionaire; his investments beginning in the wake of his lymphosarcoma had paid off.

He wrote, although editors stopped clamoring for his work: author or coauthor of an extraordinary fifty-four published pieces in 1969, his most prolific year, Lillehei saw only nine articles in print from 1974 through 1978. Lectureships dried up almost completely in the United States, although other countries, where sensibilities were different, invited Walt as often as ever. Through the end of the decade, he was visiting professor in Paris, London, Rome, Baghdad, Rio de Janeiro, Istanbul, and many other places. Lillehei accepted these invitations gratefully. They gave him a forum in which to begin the slow work of reclaiming his place in history.

Wangensteen believed Lillehei's reputation could be fully restored only if the University of Minnesota reappointed him to the faculty—in essence, pardoned him. But the university wouldn't even accept Lillehei's offer to lecture occasionally, for free; John Najarian, Wangensteen's successor, could never forget a red rose on an empty floor.

Still, near the end of his life now, Owen set off on one last crusade for his dear Walt.

He died, of a heart attack, before succeeding.

———

John W. Kirklin was president of the American Association for Thoracic Surgery when it met in the spring of 1979, in

Boston. Twenty-five years exactly had passed since Lillehei, in Montreal at the association's annual convention, had astonished the audience with his report on Pamela Schmidt. Kirklin had been a relative unknown then, still struggling in a Mayo Clinic basement to perfect a heart-lung machine.

Now Kirklin was arguably the best practicing open heart surgeon anywhere. He was one of cardiac surgery's most accomplished researchers, a scientist whose mind was sometimes compared, flatteringly, to a computer. It was no wonder Kirklin was president of the American Thoracic.

Kirklin was still straight-laced—still the bookworm holed up in the library while Walt and the boys were getting juiced at Mitch's. Kirklin did not show slides of young patients in cowboy suits when he gave a presentation; he showed only graphs and tables, confirming the results.

And yet, Kirklin was one of the very few doctors who'd publicly supported Lillehei during the dark days (Cooley and Shumway were two others). Kirklin could not be silent. It offended his sense of justice that Lillehei, always willing to share the secrets of his success with even his most formidable competitors, had been blackballed.

On that spring day in 1979, Kirklin began his presidential address with a tribute to Lillehei.

"He always was and still is a great hero of mine, because of his enormous ability and warm friendship," said Kirklin. "It's some cruel trick of fate that there is no operation called the Lillehei operation, yet he was one of cardiac surgery's greatest innovators and did scores of 'first-time' operations." Lillehei, said Kirklin, was a genius.

Kirklin then spotted Lillehei in the audience. Walt was sitting with Kaye and Craig, one of his surgeon sons.

"Dear colleagues," said Kirklin, "may I depart from my text to ask this great and pioneering cardiac surgeon to stand to your applause. Walt Lillehei, may we see you?"

Lillehei got to his feet—to a standing ovation.

THE LILLEHEIS, SHORTLY BEFORE WALT'S DEATH

Epilogue:
The Right Stuff

AUTUMN WAS FADING in Minneapolis the weekend of October 23, 1998. North in Hibbing, where Gregory Glidden had lain buried for forty-four years, the leaves were mostly gone from the trees. Soon, winter would sweep down from Canada, burying the plains in snow. Soon, summer would be a memory.

But it was warm that Saturday when Walt Lillehei's family and friends gathered at a black-tie dinner to pay tribute. The Father of Open Heart Surgery had just turned eighty.

Many of the old boys attended: Herb Warden and Morley Cohen, who shepherded cross-circulation to success; Dick DeWall, the most glittering of diamonds in the rough; and Norm Shumway, the inventor of heart transplantation.

Ex-patients also came, including Pamela Schmidt Stacherski, the Queen of Hearts; forty-nine years old now and healthy since Lillehei healed her, she worked as a receptionist and lived outside Minneapolis with her husband of twenty-six years. Mike Shaw, whose cross-circulation donor was a stranger, was there; married with four grown children, he was an entertainment consultant and rock-and-roll band leader in

243

rural Hutchinson, Minnesota. Also attending was Bradley Mehrman, Lillehei's first successful cross-circulation patient, operated on three days before Pamela Schmidt. A husband and father in Wisconsin, Mehrman had recently received a new valve, the latest in decades of treatment for heart problems. Once a salesman, he now collected disability and ran a small auto-body shop. "The last two years have been tough," Mehrman later said, "but I don't give up."

Lillehei's own health suffered; still recovering from prostate cancer and a stroke, he tired easily. His neck bent to the right—another, more recent effect of his treatment for lymphosarcoma. The ghost of Wangensteen was near.

And Lillehei still mourned the death of his son Clark—Kaye and Walt's last-born, their baby—to an inoperable brain tumor only two years before, at the age of forty-one. Walt never cried, but the thought of Clark on his deathbed had moved him to tears. "I would gladly lie down there for him," said Walt.

But what was Walt if not a survivor? "Old age is not for sissies," he sometimes now said.

Champagne flowed as dignitaries toasted Lillehei at his tribute. Declared one: "As the author Tom Wolfe would say, he's the guy with the right stuff." Former Vice President Walter Mondale delivered the keynote address, in which he thanked Lillehei for the "millions of lives" that the surgeon, his advances, and his many trainees—and *their* trainees—had saved. No one could ever calculate the exact figure, of course, but millions did not seem an exaggeration.

The week before, the Nobel prize for medicine had been awarded. This was an honor for which Lillehei had often been nominated; once again, it had eluded him. Some speculated that his tax conviction had kept him from the prize, or perhaps a longtime rival heart surgeon who sat on the Nobel Assembly, but Lillehei wasn't convinced—many other worthy candidates, he'd once noted, went unrecognized by the assembly. In any event, he'd said, "I never lose any sleep."

Lillehei had reclaimed his reputation: he was in demand again for lectures in the United States, and the University of Minnesota had finally reappointed him professor. He had become medical director of St. Jude Medical, a Minnesota company that manufactured the world's most successful heart valve, the design for which Lillehei himself had inspired. Lillehei's place in history was secure.

Helped up by Craig and accompanied by Kaye, Walt took the dais after Mondale finished. He spoke of Gregory Glidden, Pamela Schmidt, and Mike Shaw; of DeWall's heart-lung machine, which had settled the open heart quest. He spoke from memory, and sometimes memory took a moment to come. He leaned on the podium for support. He was, after all, eighty.

But there was still power in his words, in the story of all the hearts that he had touched, and the many deaths that he had cheated. Anyone who looked into his eyes saw it: the essence of Walt Lillehei remained vital, like a summer morning.

Walt celebrated Christmas with his family, and then his health began to decline. Hospitalized in the spring of 1999, he returned home only to contract pneumonia—his old nemesis, the killer of Gregory Glidden and so many other early heart patients.

Late the night of July 5, 1999, Walt's own heart stopped, and he died quietly, at his home, with Kaye by his side.

Source Notes

With the exception of C. Walton Lillehei's eightieth birthday celebration, I witnessed none of the events recounted here. This is how I reconstructed my narrative:

I interviewed virtually everyone in this book who was alive, plus family, friends, and colleagues of most of those who were not. I tape-recorded and transcribed the majority of my interviews and, whenever possible, I reinterviewed after a period of time as a check on memory. In all, I conducted nearly two hundred interviews over a period of eight years, with most occurring in 1997, 1998, and 1999.

I also relied extensively on diaries, letters, photographs, movies, and videotapes, the scientific literature, newspaper and magazine accounts, medical records, autopsy reports, and court files.

I visited archives at the University of Minnesota, the Mayo Clinic Rochester, and the Jesse E. Edwards Registry of Cardiovascular Disease (Edwards's heart collection, now in St. Paul), among others. The *Minneapolis Star-Tribune* and the St. Paul Pioneer Press opened their files to me. I was also assisted by the National Library of Medicine, the Minnesota Historical Society, the Minneapolis Public Library, the Owen H. Wangensteen Historical Library of Biology and Medicine, Harvard University's Countway Library of Medicine, Brown University's Sciences Library, the Providence Public Library, and *The Providence Journal* library. I observed open heart surgery once at Miriam Hospital in Providence, and several times at Children's Hospital in Boston, where I saw two successive chiefs of cardiovascular surgery at work: Aldo R. Castaneda and Richard A. Jonas.

More specific notes follow. The full citations of publications referenced in these notes are in the bibliography.

INTRODUCTION

Heart surgery statistics are for the latest year, 1996, as compiled by the American Heart Association.

PROLOGUE

My sources here are the same as for chapters 6, 7, and 8, detailed below.

CHAPTER 1

Because Minnesota law prohibits release of a deceased patient's medical records without consent from next of kin, I needed to find Patty Anderson's relatives. And I had a more immediate problem (with Patty, and with several other patients): when I started, I did not even know Patty's name, for her surgeons didn't remember it almost half a century later, and the scientific literature referred to her only as "a girl of six." I was beginning to despair of ever learning her name—never mind interviewing her relatives or obtaining her records when a sympathetic pathologist showed me her autopsy report; her name had been blacked out, as the law requires, but by holding the paper to the light, I could read it.

So I went to microfilmed copies of the *Minneapolis Star-Tribune* and found, in the issue the day after Patty died, a funeral notice that included the names and address of Patty's parents; thankfully, they were from Minneapolis, not somewhere else, in which case I might never have found the funeral notice. Consulting city directories beginning with the 1951 edition, I followed Betty and Lloyd as they moved around Minneapolis—and then disappeared altogether, in 1958. Had they left town? Died? Divorced? On through the city directories I went—until, in 1972, they reappeared. At least I thought it was them, for many Betty Andersons and Lloyd Andersons have lived in Minneapolis over the years.

By 1993, this Betty and Lloyd Anderson had disappeared again. But I had a relatively recent last-known address, and I learned from a neighbor that while Betty and Lloyd had died, Betty's nephew Ron Johnson lived in a suburb. The neighbor gave me his number and when I called him, he kindly signed the records-release form, shared his memories and photographs, and put me in touch with other relatives, who were similarly helpful. As it turned out, Betty and Lloyd had indeed moved out of state for many years, and Patty was their only child.

Clarence Dennis and his work: interviews of him and Lillehei; Dennis's articles; the unpublished doctoral thesis of his longtime associate Karl E. Karlson; and an interview of Karlson's widow, Gloria Karlson.

Charles Lindbergh and Alexis Carrel: an article by Dennis, and Harris B. Shumacker Jr.'s book.

Willem J. Kolff and his work: his scientific articles; the University of Utah Health Sciences Center; several mainstream media accounts; and an unpublished February 1971 autobiography, from which his quote is taken. Kolff is perhaps best known for his work directing development of the Jarvik 7 artificial heart that surgeons implanted into retired dentist Barney Clark in 1982, the first such operation in history.

Patty Anderson's operation: her medical record, Dennis's scientific articles, and interviews of Dennis.

CHAPTER 2

Lillehei's genealogy: interviews of Lillehei and an unpublished family history: "Jens Kristian Lillehei," by Lars Lillehei, Minneapolis, 1950.

Lillehei's childhood: interviews of Lillehei; Lillehei's surviving brother, James; and classmates and early friends, including Ralph Rogers, Mary Alice Stuart, Blair Pederson, Bill Smith, and George Moore. Also, photographs of the young boy and his family provided by Walt Lillehei.

John Hunter: several biographies, cited in the bibliography. The quote I use appears in John Kobler's *The Reluctant Surgeon: A Biography of John Hunter,* p. 154.

Lillehei's haunt, Mitch's: an interview of Red Dougherty's widow, Faith, and photographs Faith lent me; an interview of Jim Toye, who owned the Parker House after Dougherty died; and correspondence with Lowell Busching, a jazz expert. Also, newspaper articles, including: "Oliver Towne at Night: A Ghost Returns," *St. Paul Dispatch,* p. 30, April 22, 1960; "Oliver Towne at Night: Memories in Jazz," *St. Paul Dispatch,* p. 19, April 3, 1964; and "Remember Mitch's? It Was THE Place," *Richfield* (Minn.) *Sun,* p. 9, July 25, 1968.

Owen Wangensteen's childhood and early career: the Wangensteen collection at the University of Minnesota Archives; the Wangensteen collection at the Owen H. Wangensteen Historical Library of Biology and Medicine; an unpublished oral memoir Wangensteen gave to the National Library of Medicine over three days in 1971 and 1972; Leonard Wilson's book; many newspaper and magazine articles; and Marion Christine Wangensteen's unpublished family history. Also, interviews of Wangensteen's surviving son, Stephen Wangensteen; his daughter, Mary Brink; his half-sister, Alvina Fossum; and the family genealogist, William Wangensteen. I also interviewed Dr. Wangensteen's second wife, the late Sally Wangensteen, in September 1992.

Lillehei's wartime experiences: Lillehei's unpublished letters to his parents and Kaye, his future wife; interviews of Lillehei; Lillehei's official army record; and the army histories cited in the bibliography, including Charles M. Wiltse's *United States Army in World War II: The Technical Services . . . ,* from which I have taken the Anzio quote (p. 272). Hippocrates's admonition was quoted in Roy Porter's book, p. 187.

Wangensteen's unusual residency program: Leonard Wilson's book, several published articles, and interviews of Lillehei and nearly all of the rest of the Wangensteen-trained doctors in this book. For specifics of Lillehei's education, I used transcripts and evaluations of Lillehei in the Wangensteen collection at the University of Minnesota Archives, as well as report cards that Lillehei showed me.

Wangensteen's research: his oral memoir; his published articles; an interview of Wangensteen by Vincent L. Gott, filmed on Sept. 19, 1974, at Johns Hopkins University; interviews of nearly all of the Wangensteen-trained doctors in this book; and Wangensteen's bibliography. Compiled by Nguyen Thi Nga, health sciences library specialist at the University of Arizona College of Medicine, the bibliography includes 851 articles and books, beginning with "On the

Significance of the Escape of Sterile Bile into the Peritoneal Cavity," published in *Annals of Surgery* (volume 84, 1926).

Ogden Nash wrote his ode to Wangensteen in 1951, as part of a tribute to the professor from doctors at Johns Hopkins University. It was published in the Spring 1978 *University of Minnesota Medical Bulletin,* and elsewhere.

CHAPTER 3

Lillehei's cancer: interviews of Walt and Kaye Lillehei; Lillehei's medical record; and the pathologists' reports, from the University of Minnesota Archives.

The Broadmoor conference: Descriptions of the Hotel Broadmoor in the 1950s are found in Elena Bertozzi-Villa's book, *Broadmoor Memories: The History of the Broadmoor* (Missoula, Montana: Pictorial Histories Publishing, 1993). Lillehei presented his "The Occurrence of Endocarditis with Valvular Deformities in Dogs with Arteriovenous Fistulas," later published in *Annals of Surgery* (vol. 132, 1950).

Wangensteen's son Bud, who died in 1988: many of the same sources as for Owen Wangensteen's childhood and early career, and also an e-mail correspondence with Bud's son Kirk W. Wangensteen, who lives in Spain; two of Bud's unpublished books, copies of which Kirk graciously provided; the unpublished memoir of Bud's second wife, Shirley Pucci; records on file at the Municipal Court, City of St. Paul; records on file at Second Judicial District Court, Ramsey County, Minnesota; and records on file at Fourth Judicial District Court, Hennepin County, Minnesota.

Wangensteen's first wife, the former Helen Carol Griffin: many of the same sources as for Wangensteen's son Bud, as well as Helen's high school and college transcripts; records on file at the Hennepin County Medical Examiner; and interviews of Lillehei, Bruce Mc-Quarrie, and George Moore.

The hemicorporectomy and other radical operations: interviews of Lillehei, Norman E. Shumway, and J. Bradley Aust; and "Mid-century Invention Recalled," by F. John Lewis.

Wangensteen's second-look operation: interviews of surgeons on Wangensteen's staff in the 1950s, including Lillehei, Shumway, and Aust; and Wangensteen's articles, notably his 1954 "An Interim Report Upon the 'Second Look' Procedure . . . ," from which the story of the sixty-year-old woman was taken.

Kaye (Lindbergh) Lillehei: interviews of her, her husband, and their son Craig; and scrapbooks dating to Kaye's high school days that Kaye has kept.

CHAPTER 4

Cardiac adventures: Newspapers indeed were filled with dramatic accounts of heart patients; *The New York Times* alone published hundreds of such stories in the late 1940s and early 1950s. These are the ones that I referenced: "Ex-Marine Carrying Bullet in His Heart Hides Fact to Avoid Losing Employment," *The New York Times,* p. 23, Dec. 7, 1948; "Frogs' Hearts Replaced: Experiments by Soviet Scientist Reported by Moscow Radio," *The New York Times,* p. 26, Oct. 11, 1948; "Heart-Shocked Boy Lives," *The New York Times,* p. 29, Jan. 13, 1953; " 'Atomic Cocktail' Drinks Aid Critical Heart Cases," *The New York Times,* p. 2, April 18, 1950. The story of the "miracle man" was recounted in "Surgeon Massages Heart to Save Man, 65, Twice 'Dead' in 4 Hours," *The New York Times,* p. 1, April 20, 1950.

Lillehei let me read his diary of his 1951 trip, and the letters he wrote to surgeons back in Minnesota.

For the history of man's notions of the heart and the earliest heart surgery, I relied on Stephen L. Johnson's book; Harris B. Shumacker Jr.'s book; and Claude S. Beck's "Wounds of the Heart: The Technic of Suture," where I found the Aristotle, Ovid, Billroth, Paget, and Rehn quotations.

Dwight Harken: Lael Wertenbaker's book, several published articles, and my interview of Harken in December 1991, shortly before he died. Harken's letter to his wife, from which I have quoted, was published in Wertenbaker's book.

Robert E. Gross: Much of my understanding of the complex Gross was reached during extensive reporting for one of my earlier books, *The Work of Human Hands.* To enrich my perceptions for this book, I interviewed Lillehei and two of Gross's pupils whom I had not reached before: John W. Kirklin and Robert L. Replogle. Gross's comments on the "limited usefulness" of other approaches is from his piece in the Sept. 25, 1952, *New England Journal of Medicine.* For further background on Gross's atrial well, I read more of his articles and the passage (Ch. 62) about the well in his monumental textbook, *The Surgery of Infancy and Childhood;* the instructions I have quoted are from pp. 880–881 of this text.

Alfred Blalock and Helen Taussig: Every history of cardiac surgery that I read included accounts of the blue-baby operation that these two doctors from Johns Hopkins University pioneered. One version I especially liked was from the Hopkins Web site: http://ww2. med.jhu.edu/medarchives/page1.htm. I also interviewed Henry T. Bahnson, who was Blalock's associate for some two decades, starting early in the blue-baby era; and two other Blalock trainees, Denton A. Cooley and James V. Maloney Jr.

Charles P. Bailey: In my estimation, Bailey was the closest in spirit to Lillehei of all the pioneering heart surgeons—an extraordinary risk-taker and innovator who horrified many in medicine, especially some of the conservative academics, yet ultimately contributed more than most to the new field. I read several accounts of Bailey in the popular press, notably the cover story on him in the March 25, 1957 issue of *Time.* I was helped immeasurably by Julio C. Davila, who knew Bailey and his work as well as any surgeon still alive; Davila consented to interviews, provided me with a copy of Bailey's unpublished recollection of his early mitral valve surgery, and was the author of the invaluable 1998 piece in *Annals of Thoracic Surgery.* I also read several of Bailey's published scientific pieces, and chapters of his 1955 textbook, *Surgery of the Heart,* which almost certainly is the first textbook entirely devoted to that subject. And I interviewed Robert S. Litwak, Bailey's chief resident in the mid-1950s.

Dorothy Eustice: I knew this young woman had been a decisive influence on Lillehei when he first mentioned her, and went on at some length, with his usual detailed memory, about her and her condition. As good as his memory was, however, Lillehei did not remember when exactly she died, nor was he any longer in touch with her family. And so, as with Patty Anderson, I had my work cut out for me. Not knowing where Dorothy had lived, I began cold-calling every Eustice (and Eustis) in the Minnesota phone books—hoping that relatives, if any survived, had not left the state. My luck was immediate. On my very first call, a distant relative answered the phone: Mary K. Eustice was married to Bradley M. Eustice, whose late father was Dorothy's brother. These Eustices kindly put me in touch with other relatives who remembered Dorothy well, including: Danny Ryan, Bob Ryan, Ione Webber, Esther D. Eustice, and Hope Dennis, Dorothy's only surviving sibling, who gave me consent to view her sister's medical records.

CHAPTER 5

Lillehei's inspiration for the azygos factor: "Experimental Cardiovascular Surgery," by A. T. Andreasen and F. Watson, *British Journal of Surgery* (volume 39, 1952).

President Truman's call to arms: "Truman Asks All Aid Heart Disease Fight," *The New York Times,* p. 13, Feb. 3, 1950.

Lillehei's lab and early research: Interviews of Lillehei, Morley Cohen, Herbert E. Warden, Richard A. DeWall, Clarence Dennis, Norman Shumway, and former *Minneapolis Tribune* reporter Victor Cohn, now a visiting fellow with the Harvard School of Public Health; photographs from the University of Minnesota archives; Lillehei's scientific articles; stories from the *Minneapolis Star* and *Tribune;* Wangensteen's oral memoir; and Leonard Engel's 1958 book, *The Operation.*

Ferdinand Sauerbruch's negative-pressure chamber, and endotracheal intubation: "The History of the Development of the Negative Differential Pressure Chamber for Thoracic Surgery," by Herbert Willy Meyer, *Journal of Thoracic Surgery* (vol. 30, 1955); "Intratracheal Insufflation," by S. J. Meltzer, *Journal of the American Medical Association* (vol. 52, 1911); *Great Ideas in the History of Surgery;* and Wangensteen's *The Rise of Surgery From Empiric Craft to Scientific Discipline.* Also, several anesthesia texts.

The anti-vivisection movement: The collected papers of Maurice B. Visscher at the University of Minnesota Archives were of tremendous help. In addition to copies of letters Visscher sent to and received from politicians, doctors, and anti-vivisectionists, this collection also includes articles from: newspapers, notably the *Star* and *Tribune; Look* ("The Great Vivisection Dog Fight," by William Manchester, June 9, 1950); and *Living Tissue,* the official publication of the New England Anti-Vivisection Society, a national leader of the movement in the 1940s and 1950s. Accompanied by Lillehei, I also toured the attic of Millard Hall in 1997.

Maurice B. Visscher: In addition to reading his archived papers and articles in the *Star* and *Tribune,* I interviewed Lillehei and relied on Wangensteen's memoir.

F. John Lewis: interviews of Lillehei, Shumway, Gilbert S. Campbell, and Lewis's widow, Ruth, who provided a copy of Lewis's unpublished memoir and essays. Also, newspaper articles of Lewis and his first open heart case, and Lewis's published scientific articles.

Lewis's description of the "imaginary" surgeon is in his article "Mid-century Invention Recalled."

Hypothermia: I interviewed Wilfred G. Bigelow, Lillehei, Shumway, and others who used or were familiar with hypothermia, and I read much of the scientific literature; Bigelow's quote is from his memoir, *Cold Hearts.* Bigelow's paper at Broadmoor was published in *Annals of Surgery,* September 1950. I found the Reverend Leo Miechalowksi's testimony at the United States Holocaust Memorial Museum's web site: http://www.ushmm.org/research/ doctors/miechalowski.htm. More background on the controversial uses of cold on patients was found in Elliot S. Valenstein's *Great and Desperate Cures* and other works cited in the bibliography.

Bailey's ASD repairs: his *Surgery of the Heart,* and his 1954 *Journal of Thoracic Surgery* piece, in which this quote appears.

Lewis's historic first open heart case: interviews of Lillehei and Mansur Taufic, who assisted Lewis; the operative record; the scientific articles by Lewis and Taufic; and the many stories that Cohn wrote, including " 'U' Doctors Halt Blood of Girl 5 Minutes to Operate in Deep Cold," *Minneapolis Tribune,* p. 1, Sept. 23, 1952. I confirmed the existence of the Farm Master watering trough in the Fall 1952 Sears Roebuck Farm Catalog, a copy of which the Sears Archives provided me. The *Tribune* editorial from which I quoted, "14 Dogs Died So She and Other Children Have a Chance to Live," appeared on the editorial page of the Sept. 23, 1952, edition.

Bigelow finally attempted his first human patient in 1953, and beginning in 1954, enjoyed continued success in open heart surgery under hypothermia.

CHAPTER 6

Morley Cohen: interviews of Cohen, Warden, Lillehei, and De-Wall; and published scientific articles on the azygos factor and cross-circulation.

Herbert E. Warden: the same sources as for Cohen.

Charles-Edouard Brown-Sequard: I found descriptions of his work in Fulton's *History of Physiology.*

John H. Gibbon: interviews of Clarence Dennis, Lillehei, Victor Cohn, and author-surgeon Harris Shumacker, Gibbon's longtime friend and associate. I read Ada Romaine-Davis's *John Gibbon and His Heart-Lung Machine,* a biography; the poem I quote was reprinted in Romaine-Davis's book. The *Life* quote was from the May 8, 1950, issue. Gibbon's quotes are from his 1954 *Minnesota Medicine* piece, which was read at Wangensteen's Sept. 1953 Symposium on Recent Advances in Cardiovascular Physiology and Surgery. Gibbon's machine used a pump designed by the young Michael DeBakey in the 1930s.

Gibbon's partnership with IBM paralleled that of a Detroit-based group of surgeons led by Forest Dewey Dodrill, who worked with General Motors to develop a heart-lung machine. With its cam shaft and twelve cylinders aligned in a V, Dodrill's machine looked uncannily like a car engine.

Among the many other surgeons who were developing heart-lung machines in the early and mid 1950s: Gross in Boston; Mario Dogliotti of Turin, Italy; Clarence Crafoord, Viking O. Bjork, and Ake Senning, all of Stockholm; J. H. Tyrer of Sydney, Australia; Frank L. Gerbode of San Francisco; Jerome H. Kay of Los Angeles; Shigura Sakakibara of Tokyo; Max Chamberlain and Adrian Kantrowitz of New York; Willem Kolff and George H. A. Clowes Jr. of Cleveland; J. Jongbloed of Utrecht, the Netherlands; William H. Sewell Jr. and William W. L. Glenn of New Haven, Connecticut; and James A. Helmsworth of Cincinnati.

The Sigmamotor T-6S pump: I interviewed Roger Hungerford, son of the late Van Hungerford, who founded the company, which was based in Middleport, New York. I also reviewed Sigmamotor sales literature from the 1950s.

The beer hose: I interviewed Ray D. Johnson, president of Mayon Plastics Inc., Hopkins, Minnesota; a chemical engineer, Johnson himself designed the beer hose that Lillehei bought. Johnson invented the name "Mayon" for his company because it sounded like *nylon,* one of the exciting new synthetic materials of the time—and also like *mayonnaise,* which his father-in-law's company, Kennedy Mayonnaise Products, made.

Paul F. Dwan: interviews of Dwan's son, Peter; Dwan's widow, Eunice; and family archivist Kathleen Dwan Rabun, who loaned me a copy of Leland Schubert's *An Incomplete History of the Family of Helen and John Dwan, 1862–1945* (Shaker Heights, Ohio: The Corinthian Press, 1973). Also, several newspaper articles and University of Minnesota publications were useful.

The Mesabi Range and Hibbing: Ed Nelson, archivist for the Iron Range Research Center in Chisolm, Minnesota, who loaned me photographs and printed materials; the Hibbing Public Library; Aubin Photo Studio & Camera, in Hibbing, which has extensive archives; and members of the Glidden family, who escorted me on my visit to Hibbing.

LaDonnah Glidden: medical records from Hibbing General Hospital and University Hospital in Minneapolis; report cards, photographs, and other documents supplied by the Glidden family; interviews of Theresa Bovee, Geraldine Eicholtz, and Shirley Spinelli, three of LaDonnah's sisters; and an interview of Tom Glidden, LaDonnah's brother. I also visited the Gliddens' old residence outside of Hibbing; the house and neighborhood are essentially unchanged since the early 1950s.

CHAPTER 7

Gregory Glidden: all of the same sources as for his sister LaDonnah, plus interviews of many of Gregory's doctors, including Lillehei, Richard L. Varco, Warden, Cohen, anesthesiologist Joseph Buckley, and pediatric cardiologist Ray Anderson. I also interviewed Sue N. Sauer and Joan Campbell, nurses during the 1950s at Variety Club Heart Hospital. Because the nursing staff at that time kept virtually hour-by-hour written accounts of their patients, I was able to write a detailed narrative of Gregory's stay.

Frances and Lyman Glidden: Both are now dead, but their children helped me depict them, as did Frances's sister and brother-in-law, Beatrice and Reno Valentino.

Variety Club Heart Hospital: interviews of Lillehei and others who worked there; the University of Minnesota Archives; and a history of the Variety Club supplied by Variety Clubs International, New York. I also visited the hospital, including the auditorium and Gregory's room, now an office.

Werner Forssmann: A translation of Forssmann's original paper on heart catheterization (published in *Klinische Wochemschrift,* vol. 8, 1929) that appeared in Larry W. Stephenson's *Heart Surgery Classics;* the characterization of Forssmann as "queer, peculiar" is from a 1956 letter one of Forssmann's colleagues wrote to Dwight Harken, also reprinted in Stephenson's book.

The Lillehei-Lewis dispute: Wangensteen's papers, and interviews of Lillehei and Aust.

Lewis's anguish at losing patients: his widow, Ruth.

Source Notes

CHAPTER 8

Heart facts, statistics, and disease: the American Heart Association, and texts cited in the bibliography.

Defective heart anatomy: The two atlases that Lillehei consulted were Maude E. Abbott's *Atlas of Congenital Cardiac Disease* and Karl Freiherr von Rokitansky's *Die Defecte der Scheidewande des Herzens.*

Jesse Edwards's heart collection: an interview of Edwards; an interview of Jack L. Titus, director, Jesse E. Edwards Registry of Cardiovascular Disease; and several of Edwards's publications, most importantly his 1956 "Anatomic and Pathologic Studies in Ventricular Septal Defect," which he coauthored with Kirklin and others.

Cecil Watson: Watson's papers at the University of Minnesota Archives; Wangensteen's papers; several stories in the *Star* and *Tribune;* Wilson's book; and interviews of Lillehei, Cohen, Warden, Aust, Dennis, Gott, Ray Anderson, and Ernest A. Reiner, Watson's chief resident in 1956. The pigeon quote is from the editorial page of the Aug. 18, 1956 *Star.*

Watson's March 25, 1954, encounter with Wangensteen: This is one of the only episodes in this book that I could not corroborate with one or more direct participants (Watson, Wangensteen, and Amberg all being dead). My source was Lillehei, whose source was his mentor Wangensteen; Lillehei's memory of Wangensteen's account was clear and unchanging during several interviews. Since Lillehei's memory on other matters was so precise—and because everyone else I interviewed who knew Watson agreed his initial opposition to cross-circulation was not only plausible but likely—I have included this scene.

Ray Amberg: the University of Minnesota Archives and the *Star* and *Tribune,* which covered Amberg extensively. The quote from the member of the House Appropriations Committee was published in the Feb. 6, 1953 *Star.*

CHAPTER 9

Atoms in the news: President Eisenhower discussed "massive instant retaliation" in a press conference on March 17, 1954; Associated Press reports on the *Lucky Dragon* appeared in newspapers on March 16 and March 21, 1954; and the description of a mushroom cloud appears on pp. 541–542 of Richard Rhodes' brilliant *Dark Sun: The Making of the Hydrogen Bomb.*

Room II and its equipment: interviews of Lillehei, Warden, Cohen, Buckley, DeWall, Dennis, and OR supervisor Genevieve A. Scholtes; I confirmed their descriptions of equipment in surgery texts of the period, cited in the bibliography.

Anesthesia: Buckley's recollections were helpful, as were the previously mentioned surgery texts. The burned-chin quote was from the "Medical News" column of the Feb. 14, 1931 *Journal of the American Medical Association;* the Brooklyn accident was recounted in *The New York Times,* p. 20, April 14, 1952.

Richard Varco: an interview and correspondence with him; the University of Minnesota Archives; and interviews of Lillehei, Warden, Cohen, Buckley, DeWall, Dennis, and Shumway.

Surgery on Gregory and his post-operative course: interviews of the surgeons, reporter Cohn, and Gregory's sisters; Gregory's vast medical record; the operative and anesthesia notes; and the many scientific articles Lillehei and his group published.

CHAPTER 10

Pamela Schmidt: interviews of Pamela, Pamela's parents, and Pamela's surgeons; Pamela's medical records; the color film of the operation that an assistant to Lillehei shot; William Peters's piece in *Cosmopolitan;* and the many stories in the *Star* and *Tribune.*

The April 30 press conference: interviews of Lillehei and Cohn; Cohn's story; and the University of Minnesota News Service press release, found at the university archives.

Press reports that I cited of the first cross-circulation operations: *Time,* pp. 68–69, May 10, 1954; *The New York Times,* p. 17, May 1, 1954; the *San Bernardino* (Calif.) *Sun,* June 25, 1954; the *Egyptian Gazette,* Cairo, May 3, 1954; and the *London Daily Mirror,* May 11, 1954. The anti-vivisectionist wrote his feelings on a clip of the *Daily Mirror* story and mailed it to Lillehei, who showed it to me.

The American Association for Thoracic Surgery's 1954 meeting in Montreal: recollections of several participants, and the discussion section of Lillehei's 1954 article in *Journal of Thoracic Surgery.*

Cross-circulation later in 1954: Lillehei meticulously followed all forty-five of these patients through their lives, compiling their case histories in several articles, including his 1986 piece in *Annals of Thoracic Surgery.*

Michael Shaw and tetralogy of Fallot: interviews of Lillehei; Lillehei's published scientific papers; interviews of Shaw and donor Howard Holtz; mainstream press reports; and Shaw's mother's diary, which Shaw photocopied and sent to me. Holtz's statement to the Associated Press was published in the *Star* on Nov. 17, 1954.

The Atlantic City heckler: my interview of Sally Wangensteen, who was sitting near the heckler.

Geraldine, Dan, and Leslie Thompson: interviews of Dan Thompson; Barbara Hitt, Geraldine Thompson's sister; Gay Withers, Dan's daughter; and Lillehei and Warden. Also, photographs that Dan Thompson and Barbara Hitt loaned me, and exhaustive coverage of the later trial in the *Star* and *Tribune.*

William Mustard: interviews of Bigelow and George A. Trusler, one of Mustard's protégés; Bigelow's published papers on his monkey lung; and Marilyn Dunlop's book. Trusler also sent me photographs, Mustard's unpublished memoirs, and Trusler's own unpublished recollections, which he delivered in a 1993 lecture at the University of Ottawa Heart Institute.

Calvin Richmond: interviews of Lillehei, Gilbert Campbell, and Robert L. Vernier, the University of Minnesota pediatrician who helped arrange Richmond's trip to Minneapolis and who shared a passage of his unpublished memoir; and Campbell's papers on the dog lung. I read three stories from the *Arkansas Gazette:* "Delayed Flight," p. 1B, March 16, 1955; "Calvin Is Off to Minneapolis," p. 1, March 17, 1955; and "Wings of Mercy," p. 1B, March 17, 1955. Also, stories in the *Star* and *Tribune,* notably Cohn's "It Was Miracle Week in State Heart Surgery," in the March 27, 1955 *Tribune.*

CHAPTER 11

Richard DeWall and the bubble oxygenator: interviews of him, Lillehei, Warden, Cohen, Shumway, Dennis, Frederick S. Cross, James V. Maloney Jr., Richard A. Jonas, Robert A. Indeglia, and Howard Janneck. I read DeWall's scientific articles; every history of heart surgery also includes an account.

John W. Kirklin and his work: I interviewed Kirklin for many hours over three days in 1997 in Birmingham, Alabama, where he is professor of surgery at the University of Alabama School of Medicine; I also interviewed Lillehei and most of the same people who helped me depict Lillehei. I visited the Mayo Clinic Rochester archives and read: Kirklin's scientific articles; sales literature from Med-Science Electronics, the St. Louis, Missouri, firm that manufactured the Mayo-Gibbon heart-lung machine; and stories in the *Rochester* (Minn.) *Post-Bulletin,* notably "First 'Bypass' Operation: Girl Doing Well After Surgery," p. 1, March 23, 1955.

Earl H. Wood: My portrayal of Wood is drawn primarily from the articles he coauthored with Kirklin, and a biography written by the American Physiological Society, which I found on their web site at: http://www.faseb.org/aps/introehw.htm.

Linda (Stout) Raison: interviews of her and her father, Howard Stout, and Kirklin's written record of her operation.

James Frederick Robichaud: interviews of his mother, Catherine Robichaud, and Lillehei; Jimmy's medical record, which Mrs. Robichaud kindly gave me permission to see; and news clips and a letter from Lillehei to Catherine that another son, Pete Robichaud, photocopied and sent me.

Denton Cooley: interviews of him, DeBakey, and Lillehei; several of Cooley's scientific articles; and Cooley's book of reflections.

Michael DeBakey: interviews of him, Cooley, and Lillehei; and many of DeBakey's published pieces. As with Cooley, every history of heart surgery includes an account of DeBakey's work. DeBakey's 1955 praise for Lillehei was in the discussion section at the end of Lillehei's "The Direct-Vision Intracardiac Correction of Congenital Anomalies by Controlled Cross Circulation," *Surgery* (vol. 38, 1955).

The 1957 improved DeWall-Lillehei machine was called the sheet oxygenator, after its construction: two sheets of clear polyvinyl plastic heat-sealed to create chambers that substituted for DeWall's "mixing" tube and tubing. It was sold by Baxter Laboratories, the medical-supply firm that had participated in its development.

Eventually, Clarence Dennis and his group at the Downstate Medical Center in Brooklyn, New York, developed a practical heart-lung machine—but not without further difficulties. Working at first from a makeshift laboratory inside a former funeral parlor, Dennis built a machine that made a successful human debut in June of 1955. Dennis's next case, however, ended in tragedy when someone in the crowd of doctors visiting the operating room accidentally—and unknowingly—unplugged the machine; by the time

Dennis found the cause of the problem, his patient was dead. After that, Dennis arranged for a policeman to stand guard outside his operating room door, and went on to a long series of successful open heart operations.

Frederick S. Cross and Earl B. Kay: interviews of Cross, Maloney, Lillehei, and DeWall; an operational manual from Pemco, the Cleveland, Ohio, firm that made the Kay-Cross machine; scientific articles coauthored by Cross and Kay; and photographs and other materials that Cross loaned to me.

Cooley's approval of the DeWall-Lillehei concept was given in December of 1956 at a meeting of the Southern Surgical Association in Boca Raton, Florida, and published in Cooley and DeBakey's "Temporary Extracorporeal Circulation in the Surgical Treatment of Cardiac and Aortic Disease," *Annals of Surgery* (vol. 145, 1957).

I tracked Lillehei's operations through his scientific articles, and the unpublished case-by-case log that he kept.

The success of DeWall's bubble oxygenator soon put him in demand as a guest lecturer at medical schools and scientific meetings in the U.S. and abroad. Unable to refuse DeWall a formal residency any longer, the dean of the University of Minnesota graduate school accepted DeWall, who earned his master of science degree in 1961—years after his invention had become the most popular heart-lung machine in the world.

Lillehei's talk with Genevieve Scholtes: an interview of Scholtes.

CHAPTER 12

Earl Bakken and the pacemaker: interviews of Lillehei and Bakken; Lillehei's scientific articles; the official Medtronic company history, which I found on their web site at http://www.medtronic.com/corporate/early.html; and materials supplied by the Bakken Library and Museum, in Minneapolis. Black Thursday was described in "Power Failure in Area Causes Sizable Losses," *Minneapolis Tribune,* p. 1, Nov. 1, 1957.

Wangensteen's successor as chief of surgery: the University of Minnesota Archives' Wangensteen collection; and interviews of Lillehei, John S. Najarian, Robert Howard, and Robert A. Good, a member of Howard's search committee and then professor of microbiology at the University of Minnesota Medical School. Najarian's comments about "the loss of one man" were published in

"Prepared for Lillehei, Staff Loss, Dean Says," *Minneapolis Star,* Oct. 13, 1967.

CHAPTER 13

Christiaan Barnard and his heart transplant: My primary sources were a May 1999 interview with him, his scientific articles, and the detailed recollections he gave in *History of Transplantation: Thirty-five Recollections* and in his memoir, *The Second Life;* his descriptions of his affairs that I have quoted are from this memoir, p. 31 and p. 87. Beginning with *The New York Times's* first-day account of Barnard's historic operation, "Heart Transplant Keeps Man Alive in South Africa," p. 1. Dec. 14, 1967, I read newspaper accounts of human heart transplantation too numerous to itemize here. I also interviewed Lillehei, Shumway, and Lower.

Norman Shumway and his work: interviews of him, Lower, Lillehei, Ernest A. Reiner, and Lloyd D. MacLean; a history of Shumway's Stanford work by Eugene Dong Jr., found at http://www.stanford.edu/%7egenedong/httx/harttx.htm; Shumway's scientific articles; the June 28, 1987, *Newsday Magazine* piece; and several stories in 1968 and 1969 in *The New York Times.*

Richard Lower: interviews of him and Shumway; Lower's published papers; the June 28, 1987, *Newsday Magazine* piece.

The Chicago Medical School experiments with puppy hearts was recounted in Emanuel Marcus's 1951 article in *Surgical Forum.*

The Mississippi chimpanzee-into-human heart transplant: Although not among the great pioneers of heart surgery, James D. Hardy and his group at the University of Mississippi began to experiment with animal-heart transplantation at about the same time as Shumway and Lower. They initially intended to transplant a human heart into a sixty-eight-year-old heart-diseased man who was near death but the intended donor, a brain-dead man, lingered in a coma and Hardy had no intention of disconnecting his life support. "For a [transplant] to succeed," wrote Hardy, "the donor and the recipient must 'die' at almost the same time; although this might occur, the chances that both prospective donor and prospective recipient would enter fatal collapse simultaneously were very slim." When the sixty-eight-year-old man went into terminal shock, Hardy decided to use the heart of a ninety-six-pound chimpanzee. Although members of the transplant team had agreed not to acknowledge the operation, except at a national medical meeting, word leaked out of the University of Mississippi—but the leak was wrong. Faced with news reports that Hardy had transplanted a human heart, he was forced to tell reporters the true story. Hardy published his case in *Journal of the American Medical Association* on June 29, 1964.

Volunteers for Eisenhower: "Outlook Guarded for Eisenhower," *The New York Times,* p. 1, Aug. 21, 1968.

Shumway investigation: "Ex-Clerk on Coast Gets a New Heart," *The New York Times,* p. 7, Aug. 24, 1968. Shumway was never charged with wrongdoing.

The wrongful-death suit against Lower: my interview of Lower; the 1975 *San Diego Law Review* piece, from which Fletcher's remarks were taken; and coverage in the *Richmond* (Virginia) *Times-Dispatch,* including "Medical Definition of Death Upheld in Transplant Case," p. 1, May 26, 1972, from which L. Douglas Wilder's remarks are taken, and "Doctor Finds Trial Reinforces His Faith," p. 1, May 27, 1972, in which Lower's reflections appear.

CHAPTER 14

Beth McDonough: Her story was told to me by her mother, Mary L. McDonough, of Bridgewater, New Jersey.

Lillehei's transplants: "Dr. Lillehei Heart Graft Patient Dies: 8-Hour Operation Fails as Man Succumbs on Table," *Minneapolis Tribune,* p. 1, June 2, 1968; "Heart, Both Kidneys of Man Transplanted," Associated Press story in the *Tribune,* Jan. 2, 1969; "Lillehei Performs Heart Transplant," *Tribune,* Jan. 6, 1969; "6 Recipients Get Donor's Organs," AP story in the *Tribune,* p. 1, Feb. 21, 1969; "Lillehei Team Implants Heart in Woman, 40," AP story in the *Tribune,* April 4, 1969; "Four Organs Transplanted in N.Y. Surgery," AP story in the *Tribune,* May 14, 1969; "Teen-aged Girl's Heart and Liver Are Transplanted," AP story in the *Tribune,* July 27, 1969; " 'Christmas Present' Operation,' " AP story in the *Minneapolis Star,* Dec. 26, 1969; and "Implanting a Heart and Both Lungs: Still the Impossible Dream?" *Medical World News* (Jan. 30, 1970), from which the dialogue with the patient, forty-three-year-old Edward Falk, was taken. Also, interviews of Lillehei and Shumway.

Lillehei's offer to operate on twin boys' dog: the Associated Press, which moved stories on Jan. 3 and Jan. 4, 1969, and a captioned photograph on Jan. 4, 1969.

The Lillehei's house fire and boating accident: interviews of Lillehei and his wife, Kaye.

Lillehei in New York: interviews of Walt and Kaye Lillehei, Paul Ebert, Robert Ellis, Randy Ferlic, and two doctors then at the New York Hospital-Cornell Medical Center who agreed to speak on the condition that they remain anonymous; Wangensteen's papers at the University of Minnesota Archives, which meticulously chronicle Lillehei's troubles in dozens of letters to and from officials at the New York Hospital-Cornell Medical Center; and Adele A. Lerner, archivist for New York Hospital-Cornell Medical Center. Lillehei's comments comparing hospital tunnels to the sewers of Paris appeared in "Transplant MDs Raced Against Time," the *New York Post,* p. 1, Feb. 21, 1969, and also in "5 Get Organs from One Donor in a Series of Transplants Here," *The New York Times,* p. 1, Feb. 21, 1969. Owen Wangensteen's remark about Lillehei's painting of a nude was recalled by Owen's son Stephen in my interview of him.

CHAPTER 15

Lillehei's trial: I read the extensive record at U.S. District Court, St. Paul, Minnesota. I also read the coverage in the *Star, Tribune, St. Paul Dispatch,* and *The New York Times,* and interviewed Walt and Kaye Lillehei and Jerry Simon. The account of Lillehei's sentencing is thanks to court reporter Bruce Tiffany, who found long-lost stenographer's notes (they were missing from the official record) and then kindly transcribed them for me; without Tiffany's help for this pivotal scene, I would have been reduced to using short quotes that appeared in the newspapers.

Judge Philip Neville: stories in the *Star* and *Tribune;* "Memorial Service for the Honorable Philip Neville," Chief Judge Edward J. Devitt presiding, May 3, 1974, Minneapolis, as officially entered into the court record; and an interview with Neville's son, James, an attorney who worked with his father.

The Minnesota State Board of Medical Examiners suspended Lillehei's license to practice medicine on March 1, 1974, prompting Lillehei to write to Wangensteen: "I'm astonished at the severity of this verdict!" Lillehei's Minnesota license was returned to him a year later.

As of early 1999, the American College of Surgeons had not reinstated Lillehei.

Wangensteen's post-trial campaign for Lillehei: Wangensteen's papers at the University of Minnesota Archives.

Second International Symposium on Cardiac Surgery: Julio C. Davila's 1977 *Second Henry Ford Hospital International Symposium on Cardiac Surgery,* and an interview of Davila.

After a hearing on March 22, 1974, the state of New York declined to take action against Lillehei's license to practice in New York. Lillehei's case was bolstered by the written testimony of Denton Cooley, who said: "My debt to him will never be repaid," and by written testimony of Norman Shumway, who said, "There can be no doubt that Dr. C. Walton Lillehei is the father of cardiac surgery as we know it today."

Kirklin's 1979 tribute to Lillehei: interviews of Kirklin and Lillehei; and Kirklin's verbatim remarks, reprinted in his "A Letter to Helen," published in 1979 in the *Journal of Thoracic and Cardiovascular Surgery.*

After the Senate Watergate Committee released Richard Nixon's infamous Enemies List, in the summer of 1973, Lillehei suspected that he, too, had been targeted by the president in retribution for his public support of Humphrey and Muskie in the 1968 election. Lillehei's suspicion was bolstered by the fact that Michael DeBakey's name was on the list. Lillehei filed suit to see if he was on some as yet undisclosed list, but he later dropped the suit; no evidence emerged that Lillehei had, in fact, been punished for his political position. Nor could I find such evidence.

EPILOGUE

I attended all three days of Lillehei's eightieth-birthday celebration in Minneapolis in October 1998.

I learned of today's status of Pam Schmidt Stacherski, Bradley Mehrman, and Mike Shaw by interviewing them.

Minnesota's biomedical industry—which Lillehei's many medical contributions helped to establish—is known as Medical Alley. I learned of its economic importance from the University of Minnesota's William Hoffman; Medical Alley's web site, http://www. mbbnet.umn.edu/company_folder/ma.html; Medtronic; and St. Jude Medical.

Bibliography

GENERAL BACKGROUND ON CARDIAC SURGERY AND CARDIOLOGY

American Heart Association, web site: http://www.american-heart.org/newhome.html

Arensman, Robert M., and J. Devn Cornish, *Extracorporeal Life Support.* Boston: Blackwell Scientific Publications, 1993.

Braunwald, Eugene, *Heart Disease: A Textbook of Cardiovascular Medicine.* Philadelphia: Saunders, 1984.

Castaneda, Aldo R., Richard A. Jonas, et al., *Cardiac Surgery of the Neonate and Infant.* Philadelphia: Saunders, 1994.

Jonas and Martin J. Elliott, eds., *Cardiopulmonary Bypass in Neonates, Infants and Young Children.* Oxford: Butterworth-Heinemann, 1994.

Kirklin, John W., and Brian G. Barratt-Boyes, *Cardiac Surgery: Morphology, Diagnostic Criteria, Natural History, Techniques, Results, and Indications.* New York: Churchill Livingstone, 1986.

Netter, Frank H., *Atlas of Human Anatomy.* Summit, N.J.: Ciba-Geigy, 1989.

Perloff, Joseph K., *The Clinical Recognition of Congenital Heart Disease.* Philadelphia: Saunders, 1970.

THE HISTORY OF CARDIAC SURGERY AND CARDIOLOGY

Abbott, Maude E., *Atlas of Congenital Cardiac Disease.* New York: American Heart Association, 1936.

Bing, Richard J., *Cardiology: The Evolution of the Science and the Art.* Chur, Switzerland: Harwood Academic Publishers, 1992.

Comroe, Julius H. Jr., *Exploring the Heart: Discoveries in Heart Disease and High Blood Pressure.* New York: Norton, 1983.

Davila, Julio C., *Second Henry Ford Hospital International Symposium on Cardiac Surgery.* New York: Appleton-Century-Crofts, 1977.

Dressler, William, *Clinical Cardiology: With Special Reference to Bedside Diagnosis.* Paul B. Hoeber, 1942.

Dry, Thomas J., *A Manual of Cardiology.* Philadelphia: Saunders, 1950.

Engel, Leonard. *The Operation: A Minute-by-minute Account of a Heart Operation—and the Story of Medicine and Surgery That Led Up to It.* New York: McGraw-Hill, 1958.

Evans, William, *Cardiology.* New York: Paul B. Hoeber, 1948.

Fye, W. Bruce, *American Cardiology: The History of a Specialty and Its College.* Baltimore: The Johns Hopkins University Press, 1996.

Goldberger, Emanuel, *Heart Disease: Its Diagnosis and Treatment.* Philadelphia: Lea & Febiger, 1955.

Johnson, Stephen L., *The History of Cardiac Surgery, 1896–1955.* Baltimore: The Johns Hopkins Press, 1970.

Keynes, Geoffrey, *The Life of William Harvey.* Oxford: Oxford University Press, 1966.

Lam, Conrad R., ed., *Cardiovascular Surgery: Studies in Physiology, Diagnosis and Techniques. Proceedings of the Symposium Held at Henry Ford Hospital, Detroit, Michigan, March 1955.* Philadelphia: Saunders, 1955.

Liss, Ronald Sandor, *The History of Heart Surgery in the United States (1938–1960).* Zurich: Juris Verlag, 1967.

Moore, William W., *Fighting for Life: The Story of the American Heart Association, 1911–1975.* New York: American Heart Association, 1983.

Naef, Andreas Paul, *The Story of Thoracic Surgery: Milestones and Pioneers.* Lewiston, N.Y.: Hogrefe & Huber, 1990.

Nolen, William A., "A Short History of Heart Surgery," *American Heritage,* vol. 34, 1983.

Bibliography

Paul, Oglesby, *Take Heart: The Life and Prescription for Living of Dr. Paul Dudley White*. Boston: Distributed by the Harvard University Press for the Francis A. Countway Library of Medicine, 1986.

Porter, Roy, *The Greatest Benefit to Mankind: A Medical History of Humanity*. New York: Norton, 1998.

Rodriguez, Jorge A., *An Atlas of Cardiac Surgery*. Philadelphia: Saunders, 1957.

Senning, Ake, "Developments in Cardiac Surgery in Stockholm During the Mid and Late 1950s," *Journal of Thoracic and Cardiovascular Surgery*, vol. 98, no. 5, Nov. 1989.

Shumacker, Harris B. Jr., *The Evolution of Cardiac Surgery*. Bloomington: Indiana University Press, 1992.

Stephenson, Larry W., and Renato Ruggiero, eds., *Heart Surgery Classics*. Boston: Adams, 1994. This contains reprints and abstracts of most of the classic articles in the field.

Thorwald, Jurgen, *The Triumph of Surgery*. New York: Pantheon, 1957.

von Rokitansky, Karl Freiherr, *Die Defecte der Scheidewande des Herzens*. Vienna: Braumuller, 1875.

Wangensteen, Owen H., and Sarah D. Wangensteen, *The Rise of Surgery: From Empiric Craft to Scientific Discipline*. Minneapolis: University of Minnesota Press, 1978.

Weisse, Allen B., *Conversations in Medicine: The Story of Twentieth-century American Medicine in the Words of Those Who Created It*. New York: New York University Press, 1984.

Wertenbaker, Lael, *To Mend the Heart: The Dramatic Story of Cardiac Surgery and Its Pioneers*. New York: Viking, 1980.

Westaby, Stephen, *Landmarks in Cardiac Surgery*. Oxford: Isis Medical Media, 1997.

SURGERY AND ANESTHESIA IN THE 1940s AND 1950s

Clowes, George H. A. Jr., Robert H. Whittlesey, et al., "A Study of the Hemodynamics of Experimental Interventricular Septal Defects," *Surgical Forum*, vol. 3, 1952.

Cole, Frank, "Explosions in Anesthesia: A review of the literature," *Surgery*, vol. 18, no. 1, July 1945.

Davis, Loyal, ed., *Textbook of Surgery. Sixth edition*. Philadelphia: Saunders, 1956.

Davison, M. H. Armstrong, *The Evolution of Anaesthesia*. Baltimore: Williams & Wilkins, 1965.

Hale, Donald E., ed., *Anesthesiology by Forty American Authors*. Philadelphia: Davis, 1954.

Jones, G. W., and G. J. Thomas, "The Explosion Hazards of Ether-Nitrous Oxide-Oxygen Mixtures," *Anesthesia and Analgesia*, July–August 1943.

Keown, Kenneth K., *Anesthesia for Surgery of the Heart*. Springfield, Ill.: Charles C. Thomas, 1956.

LéeRoy, Arthur, and Brian C. Sword, "Fires, Explosions and Anesthetics," *Anesthesia and Analgesia*, July–August 1942.

Unknown, "Killed by Anesthetic Explosion," *Pennsylvania Medical Journal*, June 1940.

Woodbridge, Philip D., and J. Warren Horton, "Prevention of Ignition of Anesthetic Gases by Static Spark," *Journal of the American Medical Association*, Aug. 26, 1939.

HYPOTHERMIA

Fay, Temple, and Gerald W. Smith, "Observations on Reflex Responses During Prolonged Periods of Human Refrigeration," *Archives of Neurology and Psychiatry*, vol. 45, 1941.

Gerster, John C., ed., "General Crymotherapy: A symposium," *Bulletin of the New York Academy of Medicine*, vol. 16, 1940.

Hamilton, James B., "The Effect of Hypothermic States Upon Reflex and Central Nervous System Activity," *Yale Journal of Biology and Medicine*, vol. 9, 1937.

Kellogg, Theodore H., *A Text-Book on Mental Diseases: For the Use of Students and Practitioners of Medicine*. New York: William Wood, 1897.

Swan, Henry, and Irvin Zeavin, "Cessation of Circulation in General Hypothermia: Technics of intracardiac surgery under direct vision," *Annals of Surgery*, vol. 139, no. 4, April 1954.

Swan, Zeavin, and S. Gilbert Blount Jr., "Surgery by Direct Vision in the Open Heart During Hypothermia," *Journal of the American Medical Association*, Nov. 21, 1953.

Talbott, John H., "The Physiologic and Therapeutic Effects of Hypothermia," *New England Journal of Medicine*, Feb. 13, 1941.

Bibliography

Valenstein, Elliot S., *Great and Desperate Cures: The Rise and Decline of Psychosurgery and Other Radical Treatments for Mental Illness.* New York: Basic Books, 1986.

Virtue, Robert W., *Hypothermic Anesthesia.* Springfield, Ill.: Charles C. Thomas, 1955.

THE AZYGOS FACTOR

Andreasen, A. T., and F. Watson. "Experimental Cardiac Surgery," *The British Journal of Surgery,* vol. 39, May 1952.

Cohen, Morley, Herbert E. Warden, and C. Walton Lillehei, "Physiologic and Metabolic Changes During Autogenous Lobe Oxygenation with Total Cardiac By-Pass Employing the Azygos Flow Principle," *Surgery, Gynecology & Obstetrics,* vol. 98, 1954.

Cohen and Lillehei, "A Quantitative Study of the 'Azygos Factor' During Vena Caval Occlusion in the Dog," *Surgery, Gynecology & Obstetrics,* vol. 98, 1954.

CROSS-CIRCULATION

Bierman, Howard R., Ralph J. Byron Jr., et al., "Studies on Cross Circulation in Man," *Blood: The Journal of Hematology,* vol. 6, no. 6, June 1951.

Duncan, Garfield G., Leandro Tocantins, and Tracy D. Cuttle, "Application in Man of Method for Continuous Reciprocal Transfusion of Blood," *Proceedings of the Society for Experimental Biology and Medicine,* vol. 44, 1940.

Kerr, Edwin, Cooper Davis, et al., "The Maintenance of Circulation by Cross-Transfusion During Experimental Operations of the Open Heart," *Surgical Forum,* vol. 2, 1951.

Lillehei, C. Walton, "Controlled Cross Circulation for Direct-Vision Intracardiac Surgery: Correction of ventricular septal defects, atrioventricularis communis, and tetralogy of Fallot," *Postgraduate Medicine,* May 1955.

Lillehei, Morley Cohen, et al., "The Direct-Vision Intracardiac Correction of Congenital Anomalies by Controlled Cross Circulation," *Surgery,* vol. 38, no. 1, July 1955.

Lillehei, Cohen, et al., "Direct Vision Intracardiac Surgery: By means of controlled cross circulation or continuous arterial reservoir perfusion for correction of ventricular septal defects, atrioventricularis communis, isolated infundibular pulmonic stenosis and tetralogy of Fallot," in Lam, Conrad R., ed., *Cardiovascular Surgery: Studies in Physiology, Diagnosis and Techniques.* Philadelphia: Saunders, 1955.

Lillehei, Cohen, et al., "Direct Vision Intracardiac Surgical Correction of Congenital Heart Defects," *Archives of Surgery,* vol. 72, April 1956.

Lillehei, Richard L. Varco, et al., "The First Open-Heart Repairs of Ventricular Septal Defect, Atrioventricular Communis, and Tetralogy of Fallot Using Extracorporeal Circulation by Cross-Circulation: A 30-year Follow-up," *Annals of Thoracic Surgery,* vol. 41, no. 1, Jan. 1986.

Moller, James. H., Ceeya Patton, Varco, and Lillehei, "Late Results (30 to 35 Years) After Operative Closure of Isolated Ventricular Septal Defect from 1954 to 1960," *American Journal of Cardiology,* vol. 68, Dec. 1, 1991.

Peters, William, "A New Heart for Pamela," *Cosmopolitan,* Sept. 1954.

Prinzmetal, Myron, Ben Friedman, and Nathan Rosenthal, "Nature of Peripeheral Resistance in Arterial Hypertension," *Proceedings of the Society for Experimental Biology and Medicine,* vol. 34, 1936.

Warden, Herbert E., Morley Cohen, Raymond C. Read, and Lillehei, "Controlled Cross Circulation for Open Intracardiac Surgery," *Journal of Thoracic Surgery,* vol. 28, no. 3, Sept. 1954.

Warden, Cohen, et al., "Experimental Closure of Interventricular Septal Defects and Further Physiologic Studies on Controlled Cross Circulation," *Surgical Forum,* vol. 5, 1954.

TETRALOGY OF FALLOT

Binet, Jean-Paul, "Correction of Tetralogy of Fallot with Combined Transatrial and Pulmonary Approach," *Surgical Rounds,* Aug. 1986.

Gott, Vincent L., "C. Walton Lillehei and Total Correction of Tetralogy of Fallot," *Annals of Thoracic Surgery,* vol. 49, 1990.

Lillehei, Cohen, Warden, and Varco, "Complete Anatomical Correction of the Tetralogy of Fallot Defects: Report of a successful surgical case," *Archives of Surgery,* vol. 73, Sept. 1956.

Lillehei, Cohen, et al., "Direct Vision Intracardiac Surgical Correction of the Tetralogy of Fallot, Pentalogy of Fallot, and Pulmonary Atresia Defects," *Annals of Surgery,* vol. 142, no. 3, Sept. 1955.

Lillehei, Varco, et al., "The First Open Heart Corrections of Tetralogy of Fallot," *Annals of Surgery,* vol. 204, no. 4, Oct. 1986.

Warden, Richard A. DeWall, et al., "A Surgical-Pathologic Classification for Isolated Ventricular Septal Defects and for Those in Fallot's Tetralogy Based on Observations Made on 120 Patients During Repair Under Direct Vision," *Journal of Thoracic Surgery,* vol. 33, no. 1, Jan. 1957.

THE DEWALL-LILLEHEI BUBBLE OXYGENATOR

Clark, Richard E., Robert A. Magrath, and Thomas B. Ferguson, "Comparison of Bubble and Membrane Oxygenators in Short and Long Perfusions," *Journal of Thoracic and Cardiovascular Surgery,* vol. 78, no. 5, Nov. 1979.

Cooley, Denton A., "Recollections of Early Development and Later Trends in Cardiac Surgery," *Journal of Thoracic and Cardiovascular Surgery,* vol. 98, no. 5, Nov. 1989.

DeWall, Richard A., Herbert E. Warden, et al., "A Simple, Expendable, Artificial Oxygenator for Open Heart Surgery," *Surgical Clinics of North America,* vol. 36, no. 4, Aug. 1956.

DeWall, Theodor B. Grage, et al., "Theme and Variations on Blood Oxygenators: I. Bubble oxygenators," *Surgery,* vol. 50, no. 6, Dec. 1961.

DeWall, Warden, et al., "Total Body Perfusion for Open Cardiotomy Utilizing the Bubble Oxygenator: Physiologic Responses in Man," *Journal of Thoracic Surgery,* vol. 32, no. 5, Nov. 1956.

Gott, Vincent L., DeWall, et al., "A Self-Contained, Disposable Oxygenator of Plastic Sheet for Intracardiac Surgery: Experimental development and clinical application," *Thorax,* vol. 12, no. 1, March 1957.

Lillehei, C. Walton, DeWall, et al., "The Direct Vision Correction of Calcific Aortic Stenosis by Means of a Pump-Oxygenator and Retrograde Coronary Sinus Perfusion," *Diseases of the Chest,* vol. 30, 1956.

Lillehei, DeWall, et al., "Direct Vision Intracardiac Surgery in Man Using a Simple, Disposable Artificial Oxygenator," *Diseases of the Chest,* vol. 29, no. 1, Jan. 1956.

Lillehei, Richard L. Varco, et al., "Results in the First 2,500 Patients Undergoing Open-Heart Surgery at the University of Minnesota Medical Center," *Surgery,* vol. 62, no. 4, Oct. 1967.

Maloney, James V. Jr., William P. Longmire Jr., et al., "An Experimental and Clinical Comparison of the Bubble Dispersion and Stationary Screen Pump Oxygenators," *Surgery, Gynecology & Obstetrics,* vol. 107, Nov. 1958.

"Surgery's New Frontier," *Time,* March 25, 1957.

OTHER HEART-LUNG MACHINES AND RELATED DEVICES

Bjork, Viking O., "An Artificial Heart or Cardiopulmonary Machine," *Lancet,* Sept. 25, 1948.

Bjork, "Brain Perfusions in Dogs with Artificially Oxygenated Blood," *Acta Chirurgica Scandinavica,* vol. 96, sppl. 137, 1948.

Blum, Lester, Samuel J. Megibow, and William M. Nelson, "A Parabiotic Blood Pump," *Surgical Forum,* vol. 5, 1954.

Clowes, George H. A. Jr., William E. Neville, et al., "Factors Contributing to Success or Failure in the Use of a Pump Oxygenator for Complete By-Pass of the Heart and Lung, Experimental and Clinical," *Surgery,* vol. 36, no. 3, Sept. 1954.

Clowes, "Practical Exposure to Permit Experimental Surgical Procedures Within the Left Heart," *Surgical Forum,* vol. 2, 1951.

Dale, H. H., and E. H. J. Schuster, "A Double Perfusion-Pump," *Journal of Physiology,* vol. 64, 1928.

Dodrill, Forest D., Edward Hill, and Robert Gerisch, "Some Physiologic Aspects of the Artificial Heart Problem," *Journal of Thoracic Surgery,* vol. 24, 1952.

Gimbel, Nicholas S., and Joseph Engelberg, "An Oxygenator for Use in a Heart-Lung Appartus," *Surgical Forum,* vol. 3, 1952.

Helmsworth, James A., LeLand C. Clark Jr., et al., "A Discussion of an Oxygenator-Pump Used in Total By-Pass of the Heart and Lungs in Dogs," *Surgical Forum,* vol. 3, 1952.

Helmsworth, Clark, et al., "An Oxygenator-Pump for Use in Total By-Pass of Heart and Lungs," *Journal of Thoracic Surgery,* vol. 26, 1953.

Jongbloed, J., "The Mechanical Heart-Lung System," *Surgery, Gynecology & Obstetrics,* vol. 89, 1949.

Melrose, D. G., "A Mechanical Heart-Lung for Use in Man," *British Medical Journal,* July 11, 1953.

Bibliography

Sewell, William H. Jr., and William W. L. Glenn, "Experimental Cardiac Surgery: Observation of the action of a pump designed to shunt the venous blood past the right heart directly into the pulmonary artery," *Surgery,* vol. 28, no. 3, Sept. 1950.

Wesolowski, Sigmund A., and C. Stuart Welch, "A Pump Mechanism for Artificial Maintenance of the Circulation," *Surgical Forum,* vol. 1, 1950.

"Carrell Says 'Glass Heart' Keeps Body Tissues Alive, Asserts Possibilities Are 'Unlimited,' Hints That Children May Be Developed Artificially," Associated Press story published in the April 22, 1936, *Providence Journal.*

"The Michigan Heart," *Time,* Oct. 27, 1952.

THE PACEMAKER AND THE ROLE OF ELECTRICITY IN MEDICINE

Garner, Louis E. Jr., "Five New Jobs for Two Transistors: You'll find many practical uses for these simple, low-cost circuits," *Popular Electronics,* April 1956.

Harken, Dwight E., "Pacemakers, Past-Makers, and the Paced: An Informal History from A to Z (Aldini to Zoll)," *Biomedical Instrumentation & Technology,* July/August 1991.

Jeffrey, Kirk, "Many Paths to the Pacemaker," *American Heritage of Invention & Technology,* vol. 12, no. 4, Spring 1997.

Kirklin, John W., Harry G. Harshbarger, et al., "Surgical Correction of Ventricular Septal Defect: Anatomic and technical considerations," *Journal of Thoracic Surgery,* vol. 33, no. 1, Jan. 1957.

Levy, Morris J., Robert D. Sellers, and C. Walton Lillehei, "Cardiac Fibrillation-Defibrillation: Use of electrical current in conversion of cardiac rhythm—methods and results," *The American Journal of Medical Electronics,* October–December, 1964.

Lillehei, C. Walton, Pedro G. Lavadia, et al., "Four Years' Experience with External Cardiac Resuscitation," *Journal of the American Medical Association,* Aug. 23, 1965.

Lillehei, Vincent L. Gott, et al., "Transistor Pacemaker for Treatment of Complete Atrioventricular Dissociation," *Journal of the American Medical Association,* April 30, 1960.

Rowbottom, Margaret, and Charles Susskind, *Electricity and Medicine: History of Their Interaction.* San Francisco: San Francisco Press, 1984.

Schechter, David C., *Exploring the Origins of Electrical Cardiac Stimulation*. Minneapolis: Medtronic, 1983.

Sellers, Robert D., Jacinto Reventos, et al., "Cardiac Arrest Due to Local Use of Hydrogen Peroxide with Experimental Study of Effects of Topical Antibacterial Agents Upon the Heart," *Journal of Thoracic and Cardiovascular Surgery*, vol. 44, no. 1, July 1962.

Weirich, William L., Gott, and Lillehei, "The Treatment of Complete Heart Block by the Combined Use of a Myocardial Electrode and an Artificial Pacemaker," *Surgical Forum*, vol. 8, 1957.

Zoll, Paul M., Arthur J. Linenthal, et al., "External Electric Stimulation of the Heart in Cardiac Arrest," *Archives of Internal Medicine*, vol. 96, Nov. 1955.

HEART TRANSPLANTATION

Barnard, Christiaan N., with Chris Brewer, *The Second Life: Memoirs.* Cape Town: Vlaeberg, 1993.

Converse, Ronald, "But When Did He Die?: Tucker v. Lower and the Brain-Death Concept," *San Diego Law Review*, vol. 12, 1975.

Cooley, Denton A., "Human Heart Transplantation: Experience with twelve cases," *American Journal of Cardiology*, vol. 22, Dec. 1968.

Dong, Eugene Jr., Edward J. Hurley, Richard R. Lower, and Norman E. Shumway, "Performance of the Heart Two Years After Autotransplantation," *Surgery*, vol. 56, no. 1, July 1964.

Hardy, James D., and Carlos M. Chavez, "The First Heart Transplant in Man: Developmental animal investigations with analysis of the 1964 case in the light of current clinical experience," *American Journal of Cardiology*, vol. 22, Dec. 1968.

Hardy, Chavez, et al., "Heart Transplantation in Man," *Journal of the American Medical Association*, June 29, 1964.

Kantrowitz, Adrian, Jordan D. Haller, et al., "Transplantation of the Heart in an Infant and an Adult," *American Journal of Cardiology*, vol. 22, Dec. 1968.

Lower, Richard R., Hermes A. Kontos, et al., "Experiences in Heart Transplantation: Technic, Physiology and Rejection," *American Journal of Cardiology*, vol. 22, Dec. 1968.

Lower, Dong, and Shumway, "Long-Term Survival of Cardiac Homografts," *Surgery*, vol. 58, no. 1, July 1965.

Lower and Shumway, "Studies on Orthotopic Homotransplantation of the Canine Heart," *Surgical Forum,* vol. 11, 1960.

Lower, Shumway, and Dong, "Suppression of Rejection Crises in the Cardiac Homograft," *Annals of Thoracic Surgery,* vol. 1, no. 5, Sept. 1965.

Mann, Frank C., James T. Priestley, et al., "Transplantation of the Intact Mammalian Heart," *Archives of Surgery,* vol. 26, 1933.

Marcus, Emanuel, Samuel N. T. Wong, and Aldo A. Luisada, "Homologous Heart Grafts: Transplantation of the Heart in Dogs," *Surgical Forum,* vol. 2, 1951.

Reitz, Bruce A., Stuart W. Jamieson, John L. Pennock, and Shumway, "Heart and Lung Transplantation: Autotransplantation and allotransplantation in primates with extended survival," *Journal of Thoracic and Cardiovascular Surgery,* vol. 80, no. 3, Sept. 1980.

Shumway, Lower, and Raymond C. Stoffer, "Selective Hypothermia of the Heart in Anoxic Cardiac Arrest," *Surgery, Gynecology & Obstetrics,* vol. 109, 1959.

Stinson, Edward B., Dong, John S. Schroeder, Donald C. Harrison, and Shumway, "Initial Clinical Experience with Heart Transplantation," *American Journal of Cardiology,* vol. 22, Dec. 1968.

Zinman, David. "The Heart Transplant at 20," *Newsday Magazine,* June 28, 1987.

History of Transplantation: Thirty-five Recollections. Los Angeles: UCLA Tissue Typing Laboratory, 1991.

CHARLES BAILEY AND HIS WORK

Bailey, Charles P., "Cardiac Surgery Under Hypothermia," *Journal of Thoracic Surgery,* vol. 27, 1954.

Bailey, *Surgery of the Heart.* Philadelphia: Lea & Febiger, 1955.

Bailey, "The Surgical Treatment of Mitral Stenosis," unpublished memoir.

Bennet, Tom, "Dr. Charles Bailey, 82, of Marietta, Was Pioneer Heart Surgeon," *Atlanta Constitution,* Aug. 19, 1993.

Davila, Julio C., "The Birth of Intracardiac Surgery: A Semicentennial Tribute (June 10, 1948–1998)," *Annals of Thoracic Surgery,* vol. 65, 1998.

Rainer, W. Gerald, "The 50th Anniversary of Mitral Valve Surgery," *Annals of Thoracic Surgery,* vol. 65, 1998.

Bibliography

WILFRED G. BIGELOW AND HIS WORK

Bigelow, Wilfred G., "Application of Hypothermia to Cardiac Surgery," *Minnesota Medicine*, vol. 37, March 1954.

Bigelow, "Cold Hearts and Vital Lessons," *Bulletin of the American College of Surgeons*, vol. 69, no. 6, June 1984.

Bigelow, *Cold Hearts: The Story of Hypothermia and the Pacemaker in Heart Surgery.* Toronto: McClelland and Stewart, 1984.

Bigelow, "Creative Advances in Cardiovascular Surgery," a lecture presented on Sept. 24, 1996, at the International Conference on the History of Cardiovascular Surgery, Moscow.

Bigelow, J. C. Callaghan, and J. A. Hopps, "General Hypothermia for Experimental Intracardic Surgery: The use of electrophrenic respirations, an artificial pacemaker for cardiac standstill, and radio-frequency rewarming in general hypothermia," *Annals of Surgery*, vol. 132, no. 3, Sept. 1950.

Bigelow, W. K. Lindsay, and W. F. Greenwood, "Hypothermia: Its Possible Use in Cardiac Surgery: An investigation of factors governing survival in dogs at low body temperatures," *Annals of Surgery*, vol. 132, no. 5, Nov. 1950.

Bigelow, *Mysterious Heparin: The Key to Open Heart Surgery.* Scarborough, Ontario: McGraw-Hill Ryerson, 1990.

Bigelow, William T. Mustard, and J. G. Evans, "Some Physiologic Concepts of Hypothermia and Their Applications to Cardiac Surgery," *Journal of Thoracic Surgery*, vol. 28, 1954.

ALFRED BLALOCK

"The Blue Baby Operation," http://ww2.med.jhu.edu/medarchives/page1.htm, a page on the official web site of the Johns Hopkins Medical Institutions.

CHARLES-EDOUARD BROWN-SEQUARD

Fulton, John F., ed., *Selected Readings in the History of Physiology. Second Edition.* Springfield, Ill.: Charles C. Thomas, 1966.

GILBERT S. CAMPBELL

Campbell, Gilbert S., Norman W. Crisp, et al., "Total Cardiac Bypass in Humans Utilizing a Pump and Heterologous Lung Oxygenator (Dog Lungs)," *Surgery*, vol. 40, no. 2, Aug. 1956.

Campbell, Gilbert S., Robert Vernier, et al., "Traumatic Ventricular Septal Defect," *The Journal of Thoracic Surgery,* vol. 37, no. 4, April 1959.

Crisp, Norman W., Gilbert S. Campbell, and E. B. Brown Jr., "Studies on Perfusion of Human Blood Through the Isolated Dog Lung," *Surgical Forum,* vol. 6, 1955.

"Answer in a Dog's Lung," *Time,* April 4, 1955.

DENTON A. COOLEY

Cooley, Denton A., *Reflections and Observations: Essays of Denton A. Cooley.* Austin, Tex.: Eakin Press, 1984.

FREDERICK S. CROSS AND EARL B. KAY

Cross, Frederick S., Robert M. Berne, et al., "Description and Evaluation of a Rotating Disc Type Reservoir-Oxygenator," *Surgical Forum,* vol. 7, 1956.

Cross and Earle B. Kay, "Direct Vision Repair of Intracardiac Defects Utilizing a Rotating Disc Reservoir-Oxygenator," *Surgery, Gynecology & Obstetrics,* vol. 104, June 1957.

Kay, Henry A. Zimmerman, et al., "Certain Clinical Aspects of the Use of a Pump Oxygenator," *Journal of the American Medical Association,* Oct. 13, 1956.

Mendelsohn, David, Thomas N. MacKrell, et al., "Experiences Using the Pump-Oxygenator for Open Cardiac Surgery in Man," *Anesthesiology,* vol. 18, no. 2, March–April 1957.

CLARENCE DENNIS

Dennis, Clarence, "A Heart-Lung Machine for Open-Heart Operations: How it came about," *Transactions of the American Society for Artificial Internal Organs,* vol. 35, 1989.

Dennis, "Perspective in Review: One group's struggle with development of a pump-oxygenator," *Transactions of the American Society for Artificial Internal Organs,* vol. 31, 1985.

Dennis, Raymond E. Buirge, et al., "Studies in the Etiology of Acute Appendicitis," *Archives of Surgery,* vol. 40, 1940.

Karlson, Karl E., Dennis, et al., "An Oxygenator with Increased Capacity: Multiple vertical revolving cylinders," *Proceedings of the Society for Experimental Biology and Medicine,* vol. 71, 1949.

Karlson, Dennis, et al., "Pump-Oxygenator to Supplant the Heart and Lungs for Brief Periods," *Surgery,* vol. 29, no. 5, May 1951.

Newman, Melvin H., Jackson H. Stuckey, et al., "Complete and Partial Perfusion of Animals and Human Subjects with the Pump-Oxygenator," *Surgery,* vol. 38, no. 1, July 1955.

Spreng, Dwight S., Dennis, et al., "Acute Metabolic Changes Associated with Employment of a Pump-Oxygenator to Supplant the Heart and Lungs," *Surgical Forum,* vol. 3, 1952.

Wangensteen, Owen H., and Dennis, "Experimental Proof of the Obstructive Origin of Appendicitis in Man," *Annals of Surgery,* vol. 110, no. 4, Oct. 1939.

Wangensteen and Dennis, "The Production of Experimental Acute Appendicitis (with Rupture) in Higher Apes by Luminal Obstruction," *Surgery, Gynecology & Obstetrics,* vol. 70, 1940.

JESSE E. EDWARDS

Becu, Luis M., Robert S. Fontana, et al., "Anatomic and Pathologic Studies in Ventricular Septal Defect," *Circulation,* vol. 14, 1956.

Clagett, O. Theron, John W. Kirklin, and Jesse E. Edwards, "Anatomic Variations and Pathologic Changes in Coarctation of the Aorta: A study of 124 cases," *Surgery, Gynecology & Obstetrics,* vol. 98, 1954.

JOHN H. GIBBON JR.

Gibbon, John H. Jr., "Application of a Mechanical Heart and Lung Apparatus to Cardiac Surgery," *Minnesota Medicine,* vol. 37, March 1954.

Gibbon, "Artificial Maintenance of Circulation During Experimental Occlusion of the Pulmonary Artery," *Archives of Surgery,* vol. 34, 1937.

Gibbon, "The Maintenance of Flow During Experimental Occlusion of the Pulmonary Artery Followed by Survival," *Surgery, Gynecology & Obstetrics,* vol. 69, 1939.

Miller, Bernard J., "The Development of Heart Lung Machines," *Surgery, Gynecology & Obstetrics,* vol. 154, 1982.

Romaine-Davis, Ada, *John Gibbon and His Heart-Lung Machine.* Philadelphia: University of Pennsylvania Press, 1991.

Bibliography

Wagner, Frederick B. Jr, and J. Woodrow Savacool, eds., *Legend & Lore*. Philadelphia: Jefferson Medical College of Thomas Jefferson University, 1996.

"Artificial Heart: Mechanical Device Substitutes for Living Organ," *Life*, May 8, 1950.

"Historic Operation," *Time*, May 18, 1953.

"The Last Field," *Time*, Sept. 26, 1949.

ROBERT E. GROSS AND HIS WORK

Gross, Robert E., Elton Watkins Jr., et al., "A Method of Surgical Closure of Interauricular Septal Defects," *Surgery, Gynecology & Obstetrics*, vol. 96, no. 1, Jan. 1954.

Gross, *The Surgery of Infancy and Childhood*. Philadelphia: Saunders, 1953.

Gross and Watkins, "Surgical Closure of Atrial Septal Defects," *Archives of Surgery*, vol. 67, 1953.

Gross, Alfred A. Pomeranz, et al., "Surgical Closure of Defects of the Interauricular Septum by Use of an Atrial Well," *New England Journal of Medicine*, Sept. 25, 1952.

Gross, *Surgical Treatment for Abnormalities of the Heart and Great Vessels*. Springfield, Ill.: Charles C. Thomas, 1947.

Watkins and Gross, "Experiences with Surgical Repair of Atrial Septal Defects," *Journal of Thoracic Surgery*, vol. 30, 1955.

Watkins, Pomeranz, et al., "Experimental Closure of Atrial Septal Defects: Technique of an 'atrial well' operation," *Surgical Forum*, vol. 3, 1952.

JOHN HUNTER

Dobson, Jessie, *John Hunter*. London: E. & S. Livingstone, 1969.

Gloyne, S. Roodhouse, *John Hunter*. Baltimore: Williams and Wilkins, 1950.

Kobler, John, *The Reluctant Surgeon: A Biography of John Hunter*. Garden City, New York: Doubleday, 1960.

Mather, George Ritchie, *Two Great Scotsmen: The Brothers William and John Hunter*. Glasgow: J. Maclehose and Sons, 1893.

Paget, Stephen, *John Hunter: Man of Science and Surgeon*. London: T. Fisher Unwin, 1897.

Bibliography

JOHN W. KIRKLIN

Carey, John M., and John W. Kirklin, "Extended Radical Mastectomy: A Review of Its Concepts," *Proceedings of the Staff Meetings of the Mayo Clinic,* vol. 27, no. 22, Oct. 22, 1952.

Clagett, O. Theron, *General Surgery at the Mayo Clinic: 1900–1970.* Self-published, 1980.

Donald, David E., Harry G. Harshbarger, et al., "Experiences with a Heart-Lung Bypass (Gibbon Type) in the Experimental Laboratory," *Proceedings of the Staff Meetings of the Mayo Clinic,* vol. 30, no. 6, March 23, 1955.

Ellis, F. Henry Jr., John W. Kirklin, and O. Theron Clagett, "Tetralogy of Fallot," *Surgical Clinics of North America,* vol. 35, August 1955.

Jones, Richard E., "The Gibbon-Mayo Pump-Oxygenator," *IRE Transactions on Medical Electronics,* vol. ME-6, June 1959.

Jones, David E. Donald, et al., "Appartus of the Gibbon Type for Mechanical Bypass of the Heart and Lungs," *Proceedings of the Staff Meetings of the Mayo Clinic,* vol. 30, no. 6, March 23, 1955.

Kirklin, John W., "Present Evaluation and Future Potentialities of Cardiovascular Surgery," *Postgraduate Medicine,* March 1954.

Kirklin, "Some Lessons from the History of the Mayo Clinic," the Sixth Reynolds Historical Lecture, University of Alabama at Birmingham Medical Center, Feb. 22, 1985, unpublished.

Kirklin, "Surgical Treatment of Mitral Stenosis," *Proceedings of the Staff Meetings of the Mayo Clinic,* vol. 27, no. 18, Aug. 27, 1952.

Kirklin, John W., William H. Weidman, et al., "The Hemodynamic Results of Surgical Correction of Atrial Septal Defects: A report of thirty-three cases," *Circulation,* vol. 13, June 1956.

Kirklin, James W. DuShane, et al., "Intracardiac Surgery with the Aid of a Mechanical Pump-Oxygenator System (Gibbon Type): Report of eight cases," *Proceedings of the Staff Meetings of the Mayo Clinic,* vol. 30, no. 10, May 18, 1955.

Kirklin and F. Henry Ellis Jr., "Mitral Stenosis," *Surgical Clinics of North America,* vol. 35, Aug. 1955.

Kirklin, Donald, et al., "Studies in Extracorporeal Circulation: I. Applicability of Gibbon-type pump-oxygenator to human intracardiac surgery: 40 cases," *Annals of Surgery,* July 1956.

Bibliography

Levin, Manly B., Richard A. Theye, et al., "Performance of the Stationary Vertical-Screen Oxygenator (Mayo-Gibbon)," *Journal of Thoracic and Cardiovascular Surgery,* vol. 39, no. 4, April 1960.

Rusted, Ian E., Charles H. Scheifley, et al., "Guides to the Commissures in Operations Upon the Mitral Valve," *Proceedings of the Staff Meetings of the Mayo Clinic,* vol. 26, no. 16, Aug. 1, 1951.

Skates, Pam, "Surgeon John Kirklin: The Elegance of Efficiency," on the occasion of Kirklin's retirement from active surgery; University of Alabama at Birmingham Medical Center, 1989.

Wood, Earl H., "Special Technics of Value in the Cardiac Catheterization Laboratory," *Proceedings of the Staff Meetings of the Mayo Clinic,* vol. 28, no. 3, Feb. 11, 1953.

WILLEM J. KOLFF

Curry, Bill, "Utah University Spurs Creation of 'Bionic Valley,' Artificial Body Parts May Aid Thousands," *Los Angeles Times,* Nov. 10, 1985.

Hollobon, Joan, and Wallace Immen, "St. Peter Said Barney Clark Was Overdue," *Toronto Globe and Mail,* April 29, 1983.

Kolff, Willem J., *Artificial Organs.* New York: John Wiley & Sons, 1976.

Kolff, "First Clinical Experience with the Artificial Kidney," *Annals of Internal Medicine,* vol. 62, no. 3, March 1965.

McCarty, James F., " 'Granddad' of Artificial Organs, the Doctor Who Conducted Pioneering Work at Cleveland Clinic, Returns to Be Honored," *The Plain Dealer,* Sept. 27, 1996.

Scott, Ronald B., "A Great Medical Innovator in Utah Readies the Artificial Heart," *People Weekly,* Feb. 17, 1975.

Stephen, Robert L., *The Willem J. Kolff Festschrift.* Basel, Switzerland: Karger, 1984.

F. JOHN LEWIS

Lewis, F. John, and Mansur Taufic, "Closure of Atrial Septal Defects with the Aid of Hypothermia: Experimental accomplishments and the report of one successful case," *Surgery,* vol. 33, no. 1, Jan. 1953.

Lewis, John F. Perry Jr., et al., "Pulmonary Resection in Mental Patients with Tuberculosis," *Diseases of the Chest,* May 1955.

Lewis, Richard L. Varco, and Taufic, "Repair of Atrial Septal Defects in Man Under Direct Vision with the Aid of Hypothermia," *Surgery,* vol. 36, no. 3, Sept. 1954.

Lewis and Taufic, "The Repair of Experimental Interventricular Septal Defects, During Hypothermia, with a Molded Polyvinyl Sponge," *Surgery, Gynecology & Obstetrics,* vol. 100, 1955.

Lewis, Milton P. Reiser, et al., "Therapeutic Effectiveness of the Artificial Kidney," *Archives of Surgery,* vol. 65, 1952.

Niazi, Suad A., and Lewis, "Resumption of Heartbeat in Dogs After Standstill at Low Temperatures," *Surgical Forum,* vol. 5, 1954.

Shumway, Norman E., Marvin L. Gliedman, and Lewis, "Coronary Perfusion for Longer Periods of Cardiac Occlusion Under Hypothermia," *Journal of Thoracic Surgery,* vol. 30, 1955.

Shumway, Gliedman, and Lewis, "A Mechanical Pump-Oxygenator for Successful Cardiopulmonary By-Pass," *Surgery,* vol. 40, no. 5, Nov. 1956.

Taufic, *Memoirs.* Austin, Minn.: Self-published, 1995.

Taufic and Lewis, "Production and Repair of Experimental Interventricular Septal Defects Under Direct Vision with the Aid of Hypothermia," *Surgical Forum,* vol. 4, 1953.

C. WALTON LILLEHEI

Cohen, Morley, Robert N. Hammerstrom, Mitchell W. Spellman, Richard Varco, and C. Walton Lillehei, "The Tolerance of the Canine Heart to Temporary Complete Vena Caval Occlusion," *Surgical Forum,* vol. 3, 1952.

Frater, Robert W. M., *The Key to the Door: A Cardiac Surgical Anthology: C. Walton Lillehei.* London: ICR Publishers, 1998.

Hopkins, Mary H., ed., *United States Army in World War II: Special Studies, Chronology 1941–1945.* Washington: Office of the Chief of Military History, Department of the Army, 1960.

Kirklin, John W., "A Letter to Helen," *Journal of Thoracic and Cardiovascular Surgery,* vol. 78, no. 5, Nov. 1979.

Kirklin, "The Middle 1950s and C. Walton Lillehei," *Journal of Thoracic and Cardiovascular Surgery,* vol. 98, no. 5, Nov. 1989.

Lewis, F. John, "Mid-Century Invention Recalled," *Journal of Thoracic and Cardiovascular Surgery,* vol. 98, 1989.

Bibliography

Lillehei, C. Walton, "The Birth of Open Heart Surgery," a lecture presented on Sept. 24, 1996, at the International Conference on the History of Cardiovascular Surgery, Moscow.

Lillehei, "The Birth of Open-Heart Surgery: Then the golden years," *Cardiovascular Surgery*, vol. 2, no. 3, June 1994.

Lillehei, "Cardiac Surgery—Past Laurels and the Present Challenge," *Bulletin of the Minneapolis Heart Institute*, vol. 7, no. 1, Summer 1989.

Lillehei, "Historical Development of Cardiopulmonary Bypass," chapter in Gravlee, Glenn P., Richard F. Davis, and Joe R. Utley, eds., *Cardiopulmonary Bypass*. Baltimore: Williams & Wilkins, 1993.

Lillehei, "New Ideas and Their Acceptance," *Journal of Heart Valve Disease*, vol. 4, supplement II, 1995.

Lillehei, "A Personalized History of Extracorporeal Circulation," *Transactions of the American Society for Artificial Internal Organs*, vol. 28, 1982.

Lillehei, J. R. R. Bobb, and M. B. Visscher, "The Occurrence of Endocarditis with Valvular Deformities in Dogs with Ateriovenous Fistulas," *Annals of Surgery*, vol. 132, no. 4, Oct. 1950.

Lillehei, Ahmad Nakib, et al., "The Origin and Development of Three New Mechanical Valve Designs: Toroidal disc, pivoting disc, and rigid bileaflet cardiac prostheses," *Annals of Thoracic Surgery*, vol. 48, 1989.

Lillehei and Owen H. Wangensteen, "Effect of Age on Histamine-Induced Ulcer in Dogs," *Proceedings of the Society for Experimental Biology and Medicine*, vol. 68, 1948.

Lillehei and Wangensteen, "Effect of Celiac Ganglionectomy Upon Experimental Peptic Ulcer Formation," *Proceedings of the Society for Experimental Biology and Medicine*, vol. 68, 1948.

Lillehei, Ivan D. Baronofsky, and Richard L. Varco, "The Surgical Treatment of Congenital Heart Disease: Analysis of the results in 388 cases," *Bulletin of the University of Minnesota Hospitals and Minnesota Medical Foundation*, vol. 24, no. 3, Oct. 17, 1952.

Lillehei and Robert G. Carlson, "Surgical Treatment of Coronary Atherosclerosis: Transcribed narration and selected illustrative frames from the motion picture," Ives Laboratories, New York, 1971.

Taylor, Marjorie Hayen, *Broken Heart, Mended.* Columbia, S.C.: R.L. Bryan Co., 1972.

Bibliography

Wiltse, Charles M., *United States Army in World War II: The Technical Services, The Medical Department: Medical Service in the Mediterranean and Minor Theaters.* Washington: Office of the Chief of Military History, Department of the Army, 1965.

Various authors, "C. Walton Lillehei Surgical Symposium," *Journal of Thoracic and Cardiovascular Surgery,* vol. 98, no. 2, Nov. 1989.

"A Baby's Heart Is Mended: Operation joins mother's bloodstream to child," *Life,* Nov. 1, 1954.

"Milestones in Open Heart Surgery," a thirty-six-minute videotape produced by St. Jude Medical, 1993. This contains footage of some of Lillehei's historic operations.

"Operative technique for Utilization of the Lillehei-Kaster Prosthetic Heart Valve," Medical Incorporated, Minneapolis, 1972.

WILLIAM T. MUSTARD

Dunlop, Marilyn, *Bill Mustard: Surgical Pioneer.* Toronto: Dundurn Press and Hannah Institute, 1989.

Mustard, William T., and J. A. Thomson, "Clinical Experience with the Artificial Heart Lung Preparation," *Canadian Medical Association Journal,* vol. 76, no. 4, Feb. 15, 1957.

Mustard and A. L. Chute, "Experimental Intracardiac Surgery with Extracorporeal Circulation," *Surgery,* vol. 30, no. 4, Oct. 1951.

Mustard, Chute, et al., "A Surgical Approach to Transposition of the Great Vessels with Extracorporeal Circuit," *Surgery,* vol. 36, no. 1, July 1954.

BUD WANGENSTEEN

Bud left the manuscripts for at least two unpublished books, which I read. Neither manuscript is dated, although they appear to have been written in the 1970s. Bud used the pen name "Wangensteen":

The Eyewash Outlook: (A Strangely Innocent Hypothesis) Alter Sense Perceptions.

In the Holy Goes: An H Bomb Epistemology. Wangensteen wrote on the title page: "Suggested cover: Suffering Christ on cross (with Hitler facial hair, the man not the wood although this could have a suggestion of swastika) watched closely from bar table by drunken, young yet sleazy blond smoking banger."

Bibliography

OWEN H. WANGENSTEEN

Gilbertson, Victor A., and Owen H. Wangensteen, "A Summary of Thirteen Years' Experience with the Second-Look Program," *Surgery, Gynecology & Obstetrics,* April 1962.

Kelly, William D. and Wangensteen, "Experimental Studies on Total Gastrectomy," *Archives of Surgery,* vol. 69, 1954.

Kelly, Lloyd D. MacLean, et al., "A Study of Patients Following Total and Near-Total Gastrectomy," *Surgery,* vol. 35, no. 6, June 1954.

Lillehei, C. Walton, " 'I Remember the Chief': Impressions of Owen Harding Wangensteen from some of the surgeons he trained in the 'halcyon Wangensteen days' at the University of Minnesota." Roche Medical Image and Commentary, Sept. 1970.

Merendino, K. Alvin, Edward S. Judd, et al., "Influence of Caffeine on Ulcer Genesis: Experimental production of gastric ulcer in guinea pigs and cats with caffeine, together with a study of its effect upon gastric secretions in dog and man," *Surgery,* vol. 17, 1945.

Myers, J. Arthur, "Owen H. Wangensteen," *The Journal-Lancet,* vol. 87, no. 6, June 1967.

Portis, Sidney A., *Ulcers and Stomach Troubles: Their Causes and Relief.* Garden City, N.Y.: Hanover House, 1953.

Wangensteen, Marion Christine, *The Story of Myself.* Northfield, Minnesota: unpublished, 1921.

Wangensteen, Owen H., "Discussion of Further Experiences with a Cervico-Axillary-Mediastinal Dissection for Cancer of the Breast," *Annals of Surgery,* vol. 132, no. 4, Oct. 1950.

Wangensteen, "The Surgeon and the Ulcer Problem," *Minnesota Medicine,* vol. 28, Feb. 1945.

Wangensteen, "The Role of the Surgeon in the Management of Peptic Ulcer," *New England Journal of Medicine,* vol. 236, no. 6, Feb. 6, 1947.

Wangensteen, F. John Lewis, and Stuart W. Arhelger, "The Extended or Super-Radical Mastectomy for Carcinoma of the Breast," *Surgical Clinics of North America,* August 1956.

Wangensteen, F. John Lewis, et al., "An Interim Report Upon the 'Second Look' Procedure for Cancer of the Stomach, Colon, and Rectum and for 'Limited Intraperitoneal Carcinosis," *Surgery, Gynecology & Obstetrics,* vol. 99, no. 3, Sept. 1954.

Wilson, Leonard G., *Medical Revolution in Minnesota: A History of the University of Minnesota Medical School.* St. Paul: Midewiwin Press, 1989.

MISCELLANEOUS PUBLICATIONS

Augarten, Stan, *Bit by Bit: An Illustrated History of Computers.* New York: Ticknor & Fields, 1984.

Campbell-Kelly, Martin, and William Aspray, *Computer: A History of the Information Machine.* New York: Basic, 1996.

Kutler, Stanley I., *The Wars of Watergate: The Last Crisis of Richard Nixon.* New York: Knopf, 1990.

Lukas, J. Anthony, *Nightmare: The Underside of the Nixon Years.* New York: Viking Penguin, 1976.

Rainbolt, Richard, *Gold Glory.* Wayzata, Minn.: Ralph Turtinen Publishing Company, 1972. This book is a history of the University of Minnesota's football team.

Rhodes, Richard, *Dark Sun: The Making of the Hydrogen Bomb.* New York: Simon & Schuster, 1995.

Acknowledgments

During my long journey, people helped me at every turn. I am one lucky writer.

My gratitude first and foremost to the late C. Walton Lillehei, who opened his career and life to me, virtually to his dying day, and to Walt's wife, Kaye, who was similarly cooperative through so many interviews, telephone calls and requests, and five visits to Minnesota.

I could never adequately thank this book's many patients, and relatives and friends of patients; I am indebted to all of them, whose names are in the Source Notes. Special appreciation to the surviving Gliddens, notably Gregory's brother Tom, and three of his sisters: Theresa Bovee, Geraldine Eicholtz, and Shirley Spinelli. Gregory's story so moved me that I helped erect the monument the boy's grave had lacked for decades, and I penned his epitaph: "His little heart changed the world."

For additional help on Patty Anderson and her parents, thanks to: Lauren Anderson, Joyce Anderson, and Lavay Nelson.

For help on Sheryl Judge: Kathryn Kelly and Brian Cambern.

For additional help on James Robichaud: Rachel Cave, staff writer, *New Brunswick Telegraph-Journal;* and Jim Haley, editor of the *Oromocto* (New Brunswick) *Post-Gazette.*

For additional help on Calvin Richmond: James Moller, and Raleigh C. Petersen of the Central Arkansas Library System.

Thanks to all of the doctors mentioned in the Source Notes, and also: Alvin A. Bakst, Johann L. Ehrenhaft, Milton Henderson, Richard Hopkins, Robert A. Indeglia, Leland W. Jones, K. Alvin Merendino, Alexander Nadas, Russell Nelson, Lester R. Sauvage Sr., Michael Shea, Sara Shumway, Arun K. Singh, C. R. Stephen, Frederick B. Wagner Jr., and Solomon J. Zak.

And at the University of Minnesota: John M. Basgen, Richard W. Bianco, John S. Hokanson, Philip McGlave, Donald G. McQuarrie, and James Moller.

Thanks to all of the many librarians and archivists who helped me, with special gratitude to: University of Minnesota Archivist Penelope Krosch and her staff; Elaine Challacombe, curator, Wangensteen Historical Library of Biology & Medicine; Molly Craig, Countway Library of Medicine; Janet Schwarz, reference librarian, Virginia Historical Society; James E. Fogerty, head, Acquisitions and Curatorial Department, Minnesota Historical Society, and his staff; Nicole L. Babcock, secretary of the Mayo Clinic Rochester's Historical Unit; Stephen Greenberg, History of Medicine Division, National Library of Medicine; Andrea Stelljes, reference librarian for the Minnesota legislature; Frederick K. Lautzenheiser, associate archivist, the Cleveland Clinic Foundation; Dawn M. Stanford, administrator, IBM Archives; and Patrick Sim, curator of the Wood Library-Museum of Anesthesiology.

Newspaper help: Jim McCartney of the *St. Paul Pioneer Press;* Jerry Rising of the *Rochester Post-Bulletin;* and especially to *Minneapolis Star-Tribune* head librarian Bob Jansen and his staff.

For help with records: Jim Bymark; Vanessa Rico; and Ryan Davenport, the media relations coordinator at Fairview Health Services, who handled so many of my requests regarding the old University Hospital and patients operated on there. Ryan was always professional and courteous—a real delight.

For additional help on Bud Wangensteen: Jacky Rose, one of his daughters.

Hibbing and iron mining: Nancy Riesgrof of the Hibbing Public Library, and David Gardner of Cleveland Cliff Mining.

For help on the history of valves: Steven Khan, associate professor, University of California School of Medicine; Shelly Johnson, Medical Incorporated, which made the Lillehei-Kaster valve;

Demetre M. Nicoloff; St. Jude Medical founder Manny Villafana; Jim Ringdal of St. Jude; and Peter Gove, vice president of corporate relations, St. Jude.

For additional help on Earl Bakken and Medtronic: Karen V. Larson and Dick Reid.

For help on Dixieland jazz: Bob Ringwald, Karen Quick, Lowell Busching, Joyce Warshauer, and the All-Music Guide, www.allmusic.com.

Also: G. Franzen of the Karolinska Institute; Mary L. Lashomb, the Cleveland Clinic Foundation; Jane Starkey, America's Blood Centers; Al Eichelberger, Buick Club of America; Howard Janneck, first general manager of the artificial organs division of Baxter Travenol; S. V. Hendricks, University of Cape Town Alumni Relations Office; Phillipa Johnson, Public Relations Practitioner, Groote Schuur Hospital; Reinhard Putz of the University of Munich Medical School; Carolyn G. Thompson, senior executive assistant, American College of Cardiology; Lillehei's secretary in the 1950s and 1960s, Alice Beiersdorf Siewart; and Brenda Bauer, Lillehei's secretary today.

And thanks to Martin Campbell-Kelly, David Christianson, Elizabeth F. Crown, Joseph R. Curl, Donna Daun, Ann Earwood, Eric Fettmann, Brian J. Henry, Kelly Hislip, Robert A. Lundegaard, Madeline M. Massengale, Mary McNeilus, Phyllis Morrow, Henry G. Owen, Wesley Pylka, Maria Ramsay, Pat Rood, Linda Shelton, Helen Simeon, Jerry Simon, Allegra Sinclair, Stephen Topaz, Earl Ubell, Lindsay Young, Mary E. Youngkin, the Hennepin County Medical Examiner, the American Surgical Association, the American College of Cardiology, the American College of Surgeons, and the St. Paul public school system.

Thanks to Julio C. Davila, himself a heart surgery pioneer, who not only guided me in my reporting but also critiqued my manuscript—correcting some embarrassing errors, rooting out a few unnecessarily melodramatic touches, and always urging me to place Lillehei's achievements in the broader context of a science to which many doctors and researchers contributed. In surgery, a profession where objectivity can be a scarce commodity, Julio's was a balanced voice of reason that I deeply appreciated.

Great thanks to Eleanor and Hardy Hendren, who provided me the peaceful environment I needed for a major edit, and to their housekeeper, Irene Webb.

Acknowledgments

At *The Providence Journal,* many helped: publisher Howard Sutton, who allows me the opportunity to write these books; editor Tom Heslin, whose ideas and gracious support helped push me to the next level; reporters Brian Jones and Mike Stanton; transcriptionists Doreen Tracey and Ceci Arnold; Mike Delaney; and Linda Henderson, librarian.

This is my fourth book with editor Jon Karp—for my money, the best editor in New York. I have been privileged to work with him now for ten years. Also at Random House, thanks to Will Weisser, Carie Freimuth, Kate Norris, Monica Gomez, and Dennis Ambrose.

As always, Kay McCauley, my agent for longer than a decade, steered me through perilous waters—and made me laugh when a laugh was exactly what I needed. Thanks again, Kay.

Finally, my dear ones. A million thanks to my beautiful wife, Alexis, and my lovely daughter Rachel, for their unending support over the long haul; to my sweet boy, Cal, for his remarkable patience during all the months when he kept asking, "Why does a book take so long?"; and to my other daughter, the extraordinary Katy, my superb research assistant on many of my library expeditions.

Credits

Index

Index

heart block, 133, 143, 152, 184–87,
194–96
heart disease, as medicine's cause
célèbre, 69, 101
heart-lung machine, 12–13, 49, 54, 56,
60, 68, 78, 81, 151, 156, 164
Cross-Kay, 183
Dennis, 10–11, 14–17, 47, 56, 70,
89, 90, 93, 127, 150, 173
DeWall-Lillehei, 167–71, 176–84,
194–98, 205
Gibbon, 13, 86–90, 150–51, 173
Mayo-Gibbon, 173–76, 181, 183
heart surgery:
Anderson case, 10–11, 13–16, 56,
61, 68, 82, 89, 127
animal research, 5, 11, 13, 37–38,
54, 61, 67–82, 85–86, 87, 90–95,
105, 160–61, 208–9, 211, 219
atrial well procedure, 54–56, 78, 150
Bailey procedures, 57–60, 63,
77–79, 112
by Barnard, 205–7, 22
blue baby operation, 30–31, 32, 56,
76, 82, 154, 155, 156
in Britain, 67–68, 85–86
coronary bypass, 49, 218, 233
by Dennis, 10–11, 14–17, 47, 56,
70, 89, 90, 93, 127, 150, 173
federal funds, 69
first successful open heart case,
79–81, 85, 88, 111
Gibbon procedures, 13, 86–90
Glidden case, 3–7, 105–12, 116–23,
124, 125–38, 141, 144–45, 150,
187
hypothermia procedures, 60, 75–82,
88, 100, 111–13, 121, 133, 149,
160
Judge case, 16–18, 61, 79
misdiagnosis in, 15, 88, 112, 151
mitral valve operations, 32, 57–60,
63, 77, 78, 79
nineteenth century, 50–51
of 1930s, 30–31
of 1940s, 10–13, 30–32, 51–53, 57,
62, 101, 128
of 1950s, 14–18, 36–38, 47–64,
67–197
of 1960s, 207–23
of 1970s, 224–25, 239–41
Robichaud case, 177–80
Schmidt case, 140, 141–49, 150,
154, 155, 158
Thompson case, 156–58, 190,
191–94

transplants, 205–15
wartime, 51–53
See also open heart surgery; specific doctors, patients, hospitals, and procedures
heart wall, surgery on, 52–53
Hitchcock, Claude, 70
Holtz, Howard, 155
Hospital for Sick Children, Toronto,
160–61
Howard, Robert, 200, 201
Hume, David M., 213–14
Humphrey, Hubert, 73, 158, 223,
238
Hunter, John, 23, 26, 27, 29, 32, 57
hypothermia procedure, 60, 75–82, 88,
100, 111–13, 121, 133, 149,
160
use in first successful open heart
surgery, 79–81

IBM, 87
India, 199, 212
Internal Revenue Service, Lillehei's
problems with, 159, 201, 227–38,
244
Israel, 199
Isuprel, 185
Italy, 51, 185, 240
World War II, 27–29

Jackson, Ray W., 228–29
Japan, 183
Jefferson Medical College, Philadelphia,
56, 87, 173
Johns Hopkins University, 30, 56, 76,
186
Johnson, Jacqueline, 79–81
Johnson, Lyndon B., 198, 206
Judge, Sheryl L., 16–17, 18, 61, 70, 79,
129, 157

Kantrowitz, Adrian, 211
Kay, Earle B., 183
kidney(s), 134
artificial, 12
failure, 37–38, 62
transplant, 219
Kirklin, John W., 53, 172–76, 197,
222, 239, 240–41
heart-lung machine, 172–76, 181,
183
Klett, Joseph, Jr., 214
Kolff, Willem J., 12

Lasker award, 198
leukemia, 91

Index

Index

Index

G. WAYNE MILLER, a staff writer at *The Providence Journal,* has won many awards for his writing. This is Miller's fifth book. Visit him on the Internet at www.gwaynemiller.com, where readers are invited to post their own stories of heart surgery to a special *King of Hearts* bulletin board.